THE LORD IS MY SHEPHERD

THE LORD IS MY SHEPHERD

An extraordinary account of aerial
combat over Europe during WWII

John F. Wilkinson

Compiled and Edited by E. Gordon Walker

Two Geez Co.
Plymouth, MN
2016

THE LORD IS MY SHEPHERD

ISBN 978-0-9974451-0-7

This book is a factual account of events occurring over 70 years ago. Names, organizations, places and events are true and are captured on these pages as accurately as memory permits. We apologize for any errors and omissions and will endeavor to correct them in subsequent editions.

First Printing: 2016

For information contact:

Edward G. Walker
Two Geez Co.
3715 Vinewood Lane North
Plymouth, MN 55441
gordon@twogeez.com

www.twogeez.com

Dedication

This account of my personal experiences in aerial combat over Europe during World War II is dedicated to my Lord and Savior, Jesus Christ. It is without question that His hand has saved me when by all the natural laws I should have met my end. I pray that reading these experiences will serve as a means through which an interior spark is ignited bringing you closer to the Lord.

To my sister Eleanor Mary Morgan who up to the day of her death in November 2015 continually inspired me to enrich my relationship with the Lord. Eleanor and I formed a lifelong bond from our very earliest childhood. While we led vastly different lives, we always cherished opportunities to be together. She joined the Norwegian High Command in London at age 18, learned the language and went with them when they returned to Oslo, Norway. She was in Norway for five years. Along the way, MI5 contacted her and she became an operative, occasionally with near fatal consequences. Since she and I were the active members of our family in WWII and later, she liked to say, "We lived life recklessly with care!"

And to my sister Joan who served her country honorably during the war as a Radar Officer.

No writing of this nature would be proper without recognition of the thousands of men and women who gave their lives so we could enjoy freedom.

JFW

A Psalm of David

The LORD *is* my shepherd; I shall not want.

He maketh me to lie down in green pastures: he leadeth me beside the still waters.

He restoreth my soul: he leadeth me in the paths of righteousness for his name's sake.

Yea, though I walk through the valley of the shadow of death, I will fear no evil: for thou *art* with me; thy rod and thy staff they comfort me.

Thou preparest a table before me in the presence of mine enemies: thou anointest my head with oil; my cup runneth over.

Surely goodness and mercy shall follow me all the days of my life: and I will dwell in the house of the LORD for ever.

The Holy Bible, King James Version. Cambridge Edition: 1769; *King James Bible Online,* 2016. www.kingjamesbibleonline.org

Contents

Acknowledgements

I owe my solid foundation in the art of flying to flight instructors Arnold and Bing who were my mentors in the PT-17 and the AT-6 respectively. They were both United States civilians. They were extremely competent and very friendly. They both contributed greatly to my flying ability. Arnold taught me so well that in the end he became the student and I, the instructor.

41 Squadron Leader Benham was a fine friendly man during my time with the squadron. And Group Captain Johnny Johnson was a real gentleman who gave me special attention. Both these men set the standard for leadership and moral values. They contributed mightily to my development as both a fighter pilot and a man.

I must also thank my good friend Bill Nichol, editor of a monthly paper for the retired officers association in Redding, California who first urged me to write articles about my experiences.

I give very special thanks to Gordon Walker who carried the burden of the requirements to publish this book.

Finally, I thank my friend Douglas Shonley whose computer skills he so freely shared enabled progress without disaster and worldwide distribution of my gun camera film.

JFW

John F. Wilkinson

Map of Western Europe

England

- Northampton

- London
➢ Warmwell   ❖ Horsham ➢ Lympne

Legend
- Major cities
❖ Christ's Hospital
➢ 41 Squadron Locations

Map of Western Europe showing locations of 41 Squadron from mid-1944 through end-1945

Denmark

➤ Copenhagen/Kastrup

➤ Husum

➤ Lubeck

▪ **Hamburg**

➤ Celle

▪ **Berlin**

Amsterdam
➤ Twente

➤ Eindhoven
➤ Ophoven
➤ Diest/Schaffen
➤ Evere

Belgium

Germany

France

John F. Wilkinson

Part 1: The Early Years

I WAS BORN on the first day of February 1923 to Jack and Eleanor Wilkinson. They lived in a comfortable country home called Courtenhall near Northampton, England. At the time, they had a three-year-old daughter Joan. Sometime during the first few months of my life, while nursing at my mother's breast I suddenly became violently ill. I was not expected to live more than two weeks. However, owing to the hand of God, I survived. The family moved to Stone House in the village of Dallington, also near Northampton, where my father owned and operated a business manufacturing and selling quality boots and shoes. In 1925 along came another daughter, Eleanor. There were many happy days for us in our early years. My parents traveled frequently to the U.S. for both business and pleasure.

John F. Wilkinson

Winchester House School

At age eight I attended Winchester House School, Brackley, Northamptonshire, a boys boarding school preparatory to entering Eton (an upper class institution). Do you remember your history lesson about King John and the Magna Carta? Since the Norman Conquest of England, the ruling Kings sought to increase their power over the Lords and Barons who felt bridled under royal restrictions. In 1215, following King John's defeat during an incursion into France, the Lords and Barons were emboldened to present John with the Magna Carta. This was the beginning of the English form of government, which is the basis of our own here in the United States. The alternative was open revolt, something the king could no longer be assured of putting down. The night before the Magna Carta was presented to King John at Runnymede, the document spent the night in a fortified castle at Brackley, Northamptonshire.

Why the history lesson? Well it so happened that 716 years later I lived in that same castle, now Winchester House School. Newer facilities and an extension transformed the castle into a prestigious and expensive preparatory school. The castle had battlements, secret passages and a dungeon. We boys were not permitted into any of these secret places until the day we were leaving school, which I did in 1932 when I was nine. The financial crash in the U.S. precipitated the loss of my father's factory. In despair my Dad took his own life shortly thereafter and I had to leave the school in Brackley.

Christ's Hospital

After my father's death my mother obtained a school governor sponsor for me in order to be admitted to another ancient boy's boarding school, Christ's Hospital, Horsham, Sussex. Following entrance exams I was admitted to this school of considerable academic note founded by Edward VI in 1552, originally for orphans and later for boys with one parent. It was not a hospital as we know hospitals today. The dictionary defines 'hospital', among other definitions, as a place of refuge. In the sixteenth century, the plight of orphaned children was pitiful. Young King Edward VI was very much aware of this and so in 1552 he established a 'place of refuge' and learning for some of those orphans. Schooling in those days was mainly for the wealthy and royalty, so Christ's Hospital became one of the first 'public' schools. The English public school is actually what we now call a 'private' school. A group of wealthy governors runs Christ's Hospital. As the centuries passed, other civic facilities and state run schools became available for orphans, so Christ's Hospital's school governors revised their rules to accept children with one parent. Academically it was considered one of the premier schools of England and covered the span from preparatory school through college, which was beyond the level of our local colleges, in preparation for university.

Christ's Hospital was originally in London. Sometime in the late 1800's or early 1900's a very large area of land in Sussex, south of London, was acquired and the present school was built. It included playing fields, a teaching farm, a post office, a gymnasium with swimming pool, a slightly remote music facility (to keep discordant sounds under control) and its own railway station. The buildings consisted of ten 'H' shaped houses spread evenly around a curved tree-lined avenue, five on each side of a central quadrangle. Each house had its own name. I was in Middleton A, on the left side of the 'H'. The right side was Middleton B. There were 50 boys in each side of each house. The dormitories were on the second and third floors. The center of the 'H' housed the masters' offices on the ground floor and on the second floor, the matron and her assistant who oversaw the dormitory housekeeping.

The large central quadrangle had a fountain, lawns and walkways and was surrounded by the dining hall, the classrooms, the church with a fine pipe organ and the auditorium. On each of the sides save the dining hall, the walkways were covered by arched cloister-like overheads. Dinner was served at noon. At dinnertime, the boys from each house assembled in phalanxes four abreast, marched up the avenue and into the dining hall to the marching music of the school band playing in the quadrangle. One thousand boys marched into that huge hall to their specified long tables where they sat on backless benches tightly pressed shoulder to shoulder. The masters' tables were on a raised platform at one end of the hall. There was a raised pulpit against one wall in the center of the hall from which a senior boy said the mealtime 'grace'. Can you imagine the noise of 1000 chattering boys, plus masters, all in one place? One time I actually saw one of the heavy drinking glasses disintegrate in response to a scream that emanated from somewhere in the hall!

In 1552, a school uniform was designed that has changed little to this day. There were two overlapping small white cotton rectangles at the neck known as 'bands'. The silver buttons at the front of the tunic and sleeves carry the image of Edward VI. As schoolboys will do, we occasionally would put one on the nearby railway lines and it would be flattened to look like a shilling. You probably don't remember that there used to be two far-things to a ha'penny, two ha'pennies to a penny, twelve pennies to a shilling and twenty shillings to a pound. Then there were thre'penny bits, six penny bits, a florin was two shillings, a half crown was two shillings and six pence and a guinea was one pound and one shilling. For-tunately, we do not have to deal with that kind of arithmetic today!

The tunic of the navy blue serge school uniform flared out into an ankle length skirt, slit open at the front. It was hot in summer and cold in winter. Under the skirt were black knee britches with silver buttons by the knees, bright orange stock-ings and black shoes. One change for practical reasons, the shoes no longer have a big buck-le, but are now modern lace up shoes. The type of leather belt and buckle hanging loosely around the waist denoted

Age 9 in my Christ's Hospital uniform

whether you were a junior or a senior. Another little quirk in the history of the school is that the big floppy hat has been dispensed with, because one day a boy went in to see Queen Victoria and forgot to take his hat off! Hats were banned from then on. We did have thin shorts and

T-shirts for sports, which included rugby football (rugger), soccer, fives (handball), squash, cricket, hockey and lots of running and physical exercises; otherwise the uniform had to be worn at all times. Therefore, when playing outside at break times we folded the skirt of our uniforms to form a V shape behind us and then rolled it up around the belt to form a big roll at our backs leaving our legs free for running.

The school had military style discipline, with corporal and other physical punishments for infraction of the rules. Certain masters (teachers) and prefects (senior boys) dispensed some of it all too eagerly. For me, the discipline in the Royal Air Force was freedom compared with the school discipline. Together with an excellent education, I participated in other activities available. One was the Royal Officers Training Corp (ROTC) program; another was the Boy Scouts, a fine organization from which I acquired many skills beyond academic studies.

All the school's houses were joined to the dining hall by a tunnel that was only used in wet weather and quite frequently in the winter months. I was still in school when the war started and we had the job of carrying hundreds of sand bags into the tunnel to form staggered floor to ceiling blast walls for bomb shelters yet leaving access to traverse from one end to the other. When the siren sounded, we all hurried down to the tunnel. At that time I also joined the Local Defense Volunteers (LDV), later renamed the Home Guard. Our duty was to patrol a given area in shifts at night to watch for German paratroopers. Having already been trained by the ROTC in the use of firearms, we were issued a rifle and one, yes one, bullet. Since there was no means of communication, it really was an exercise in futility. However, we felt we were doing our bit. When on duty, but not actually on patrol, we would nap in a barn. The barn was full of field mice that ran all over us as we lay on the floor. However, by then we were so tired we ignored them and napped anyway.

As active schoolboys, getting enough food was always a problem. Meals were wholesome but we regarded them as inadequate. One time we presented a big petition to the staff and the meals did improve, temporarily. On Sunday afternoons, we were permitted to go for walks in the surrounding countryside. These occasions became times of foraging for anything in the fields that was remotely edible. It is surprising what you will eat when you are hungry. In the apple season, we mounted military style operations to gather fruit. This was a rather reprehensible activity, but I never heard of a complaint from any of the farmers. One event was difficult to explain. There was a church on one side of the central quadrangle that had a very high and steep slender steeple. One morning it was discovered that during the night one or more daring boys had placed a chamber pot on the top of the steeple. How they got it there in the dark was a mystery, especially since the masters could not figure how to get it down. Eventually they called the ROTC Master at Arms who shot it down!

Christ's Hospital has undergone a number of changes since I graduated. When my wife and I visited the campus in the late 1970's I was surprised to see it had become co-educational. In addition, it appeared that the wearing of the uniform was no longer as strict as it used to be. The uniform was so well-known that wearing it on specific occasions when on holiday gave one free access to places like the Tower of London and to areas of St. Paul's Cathedral not open to visitors. While visiting the school campus we stood admiring the interior of the cathedral-like church. Along each wall facing the center aisle were banks of pews enough to seat a thousand boys, masters and visitors. As we gazed about, a very small boy came walking up the long tiled aisle. In the reverent silence, the sound of the boy's shoes on the tiled floor echoed throughout the building. He climbed up the winding stairs and disappeared into the pipe organ console, which has pipes at both ends of the church. Shortly thereafter, the church was filled with the most beautiful music.

There have been a number of well-known or famous persons who have risen from the Christ's Hospital school ranks to make their mark upon the world. Some of them are memorialized with large oil paintings hanging on the dining hall walls. However, other than Coleridge and Bacon, my memory of their names has faded to the point that I cannot name them any longer. It is surely a place of interest to history buffs and teachers.

Start of WWII

At about age twelve I became seriously ill and once again was not expected to live. The Lord interceded again and although I lost six months schooling, I graduated in 1940 at age seventeen and returned to live with the family, now in London. England declared war on Germany on September 3, 1939. I was still in boarding school. Later while living at home I watched the aerial Battle of Britain from the ground. Standing in front of our home I witnessed the fighters perform their deadly ballet of condensation trails and machines falling from the sky over southeast London and to the coast. Little did I know that soon I would be doing the same thing over Germany and other countries.

The night bombing of London began soon thereafter. When Hitler and Goering realized that a cross-channel invasion of England could not succeed without mastery of the air they switched to night bombing. Our pilots and aircraft had almost reached the breaking point and the night bombing gave us a moment of respite to rearm and train more pilots. The night bombing was deadly and disastrous in terms of lives and property destroyed. The noise of the anti-aircraft guns firing and the bombs exploding throughout the night was horrendous. One night the house two doors from us was bombed and destroyed. Our windows were shattered, walls cracked and all our heavy plaster ceilings fell down. Each morning when daylight came, the streets and lawns were littered with small, very sharp shell fragments from the anti-aircraft projectiles. At the time, I had a job in the city near St. Paul's Cathedral. Just getting to work was a challenge as it meant scrambling over rubble and walking between burning buildings to get there. I never knew until I arrived whether my place of employment would still be there. The road, known as Ludgate Hill, that led up to St. Paul's was often covered with hoses. As I walked up the street, I frequently passed buildings with flames pouring from their windows. There wasn't enough water to put them out and the firemen moved to areas where they could be more effective. I recall an evening when returning from a movie theater some distance from our home in Hampstead Garden Suburb. It was quite dark and my bicycle lamp was not working. There was no traffic and I was able to see the outline of roofs on each side of the street. As I peddled my way down the middle of the street I silently brushed between two pedestrians who had the same idea. I peddled harder as I was followed by a stream of invectives!

Hooked on Flying

While visiting Sywell aerodrome, a grass field near Northampton, England I was mesmerized by a biplane. It seemed to me to be a large one with a small cabin seating four passengers. The pilot's cockpit was above and in front of the cabin. At the time, I was fifteen and still several years away from graduating from school. Nonetheless the activities and machines at the air display fascinated me. My mother bought a ticket, costing seven shillings and sixpence, for me to take a brief flight in the biplane. This was my first introduction to flight and the start of a passion for flying – I was hooked!

My company sent me to Nuneaton, Warwickshire where they had acquired a satellite facility. While at Nuneaton I reached the age of eighteen and soon thereafter boarded a bus to Coventry where I volunteered for RAF flight training.

Part II: Training

TO FLY IN THE ROYAL AIR FORCE one must volunteer. In December 1941 at the age eighteen I volunteered for flight training. I requested pilot training and was immediately accepted.

Ground Training and First Ten Hours Flying

Following the various Royal Air Force induction processes, I was sent to Cambridge and assigned to Pembroke College. Cambridge is a beautiful old university town set on the river Cam where we could enjoy punting in our very limited free time. It was, for the most part, an enjoyable time. The heavy oak doors to the college were closed at 10 p.m. However, for those missing the deadline, there was a coal chute into the basement so well used that one did not get dirty when entering later in the evening! In

Cambridge, there was a small rather decrepit movie theater that we frequented. On one occasion I stood up to let a couple leave and the whole row of seats fell over. On another, I felt something crawl up my trouser leg, I grabbed my trouser leg and signaled the attendant who had a torch (English for flashlight). In the torchlight, we saw a mouse that I killed fall out of my trouser leg. I went on to enjoy the rest of the movie.

Upon joining the Royal Air Force, and being with men mostly a bit older than myself, I realized how little I knew of the ways of the world and cautiously learned all I could. Having been in a strict boarding school, serving the crown seemed like freedom to me, while to many others it was irksome discipline. On completing and passing final exams, I was posted to Bottisham, an airfield near Cambridge. Bottisham was a grass field and there I got my introduction to the DeHavilland Tiger Moth, the most basic of biplane flying machines. There was a great deal to learn, starting with months of ground school and then ten hours of flying instruction. The object was to determine our potential to become pilots. Along with others, I was given ten hours of flight training to determine whether I had the potential to be a pilot. I soloed in 8 hours. That sealed my future.

A fighter pilot is not made over night. Before one even sets eyes on an aircraft, there are long months of ground school. The training at Pembroke College was very intense covering all the aspects of flying from complete knowledge of the mechanics and construction of aircraft to every aspect of flying, engines, aerodynamics, airframes, weapons, bullet and bomb trajectories, aircraft recognition and meteorology, radio communications,

Morse code, navigation, day and night and discipline. Then the day arrives when one takes that first flight as a student in a primitive biplane to test one's potential aptitude to become a pilot and to experience the exhilarating joy of flying. To do this alone, after only 8 hours of flying is an unforgettable personal achievement.

Riddle McKay Aero College

England was now an island fortress in the truest sense of the phrase. Every airfield was needed for operational purposes. In addition, the weather was not conducive to intensive flight training. Consequently we pilot aspirants were sent to South Africa, Canada and the U.S.A. for training. I was posted to the Aircrew Dispersal Centre Manchester on 24 June 1942 where we were housed in tents. During my two weeks there, I don't remember the rain stopping once. Toward the end of the two weeks early one day we were all lined up to receive postings. The RAF had a very unfair system not only requiring alphabetical order but also starting with the fortunate ones whose names began with the letter "A." This protocol was followed anytime a line was required - postings, pay day, what have you. Remember it was raining as usual and the "W's" were nearly the last ones to receive their postings. About the first of July I was transferred to Glasgow, Scotland where I boarded the freighter "HMS Letitia" bound for Canada. How many of us were jammed in the ship's hold I do not know, but probably well in excess of a thousand. We had the option of bedding down on the steel floor or in a hammock. Thinking a hammock to be safely above those who might puke, that was my choice. However, I found I could not turn over and had to sleep shaped by the hammock. So looking around I found someone bedded on the floor without a hammock above him. I learned he preferred a hammock so I proposed we switch, which we did.

We were so valuable to the war effort that no less than three destroyers protected our lone freighter. One night I woke up and groped my way to the head. To my astonishment when I flushed the toilet it glowed in a bright green iridescent color. I stood there for some time just flushing the toilet, which obviously used seawater. The same little sea creatures that glowed when disturbed also cause the ship's wake to glow in similar fashion that could be a dead give-away if there were any nearby enemy ships or U-boats.

About 9 July 1942 we arrived in St. John harbor where I was delighted to see one of those huge four engine passenger flying boats. It was not for us however. We all boarded Canadian trains and were taken to Moncton, New Brunswick arriving the next day. We remained there about a month.

During the intervening time, we explored the local countryside. One day two or three of us rented bicycles to go and investigate a phenomenal claim there was a place where water flowed up hill. At first sight, it really did appear to flow uphill. However on closer inspection it became obvious the local landscape and hills simply made it an optical illusion. We were subsequently assigned to airfields in Canada, California or Florida. All my friends were scheduled to go to California but for some unknown reason I was earmarked to go to Florida. I tried to get my posting changed to be with my friends, to no avail, but as it happened, the California postings were cancelled and they were all sent to Canadian airfields. Once again, the hand of the Lord was at work. I boarded a train bound for New York City and soon made new friends. In the time given to us to explore New York, some of us went to the top of the Empire State building where we were not supposed to take photographs. I had purchased a nice compact camera that I hid in a paper bag filled with bananas and so I was able to take photos from the top of the building.

The next train on which we traveled was something out of yesteryear or beyond! I had never seen such antiquated carriages. They must have come from the early times of rail travel for transporting Chinese laborers and other foreign workers. The seats were wooden slats. At the end of the carriage, there was a hole in the floor with a short wall around it. This served as our bathroom! We spent five uncomfortable days in those carriages. There were a few more brief stops and our cars were, on occasion, shunted aside for transportation that was more important. We quickly learned to disassemble the seats at night so we could lie down stretched out to a normal sleeping position.

Arriving at Clewiston, Florida, beside Lake Okeechobee, we were taken a few miles down the road to the Riddle McKay Aero College airfield and barracks where there was a British Flying Training School run by Royal Air Force officers using American flying instructors. My fellow students and I had finally arrived for flight training at the Riddle McKay Aero College in the Everglades.

Our living units consisted of two sets of rooms housing four men each with adjoining bathrooms in the middle. The drinking fountains outside smelled so strongly of sulfur and goodness knows what else, one pre-ferred not to walk near them. Other than that, we were comfortably situated with good meals and a small café where we could procure additional food while off duty. When time permitted, we would hitchhike to Palm Beach or Miami. There was a very nice hotel in Miami, the McFadam-Deauville, where we were given special rates considering our pay was 50 cents per day that included something they called danger pay.

At Riddle McKay Aero College, I undertook basic and advanced flight training. I reveled in flying and whereas some students just wandered around the skies when flying solo I spent every moment practicing every maneuver taught and some not taught. I was determined to be the best and was not about to let any Luftwaffe pilot get the better of me. In fact, I not only got my commission but also was awarded "Best Flying Cadet" with a sterling silver bracelet so noting the course number and date of training. Even today, I wear it proudly on occasion.

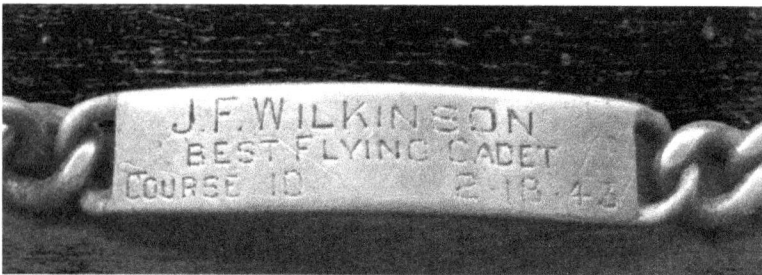

Best Flying Cadet

In Florida, we studied the many required subjects for half the days and received flying instruction the other half of each day. We had a great deal to learn. For basic train-ing we flew the Boeing Stearman Model 75 biplane, designated PT-17. I had a very good instructor, a gentleman named Arnold. I really liked him, we got on well together and I thought of him as being very old (probably 40 or 45!). I quickly became very proficient at all phases of flying, especially aerobatics, and unlike some of my fellow students, I spent every minute of solo flying diligently practicing all the maneuvers. He taught me so well that by the time the end of basic training approached he was asking me how I performed some of the maneuvers he had been teaching me but with which he had some difficulty. So in fact the student discretely became the teacher. He was an excellent instructor.

During flying training there were always little incidents that captured our attention. Flying the PT-17 the student (or pilot when flying solo) sat in the back seat and the instructor in the front seat. Communication between student and instructor was through a primitive voice tube. One day one of the instructors told the student to practice spins, which the student duly did. The spinning continued longer than it should have so the instructor told the student to pull out. No response. He shouted through the tube, but still no response. He turned around and there was no one in the back seat. Instant panic set in! The primitive lap seat belt closed with a single simple latch. The student's speaking tube became entangled with the seat belt latch unlocking it, so when he attempted to correct the spin the centrifugal force catapulted him out of the cockpit. He bounced off the tail, deployed his parachute and landed safely in the swamp below. He was safely recovered from the swamp by a swamp buggy, a vehicle with huge tires and a high platform for the engine and driver. Needless to say the instructor recovered and landed safely. It is not a good thing for an instructor to lose a student in flight!

Developing proficiency in instrument flying was the next step in training. Along with ongoing ground school, there were many hours in the dark and confined isolation of the Link trainer. This was combined with a progression of flying exercises in a series of increasingly more complex maneuvers. I spent a total of 100 hours in the Link trainer polishing blind flying capabilities and learning to have complete trust in instrument flying. While many flew aimlessly about the sky during their solo training periods, I spent every moment practicing and preparing to handle all manner of emergencies. I believed this was essential to be able to excel in combat. These exercises were, in my mind, paramount to the discipline of safe flight and the lessons learned were not to be ignored. The pilot who is dedicated to flight eventually gets to the point where the aircraft becomes an extension of his mind and body, with no thought given to the mechanics of controlling his machine as conditioned response takes over. Man and machine become one and performance is the result of reaction as opposed to thinking how to control the machine. Engine and other instruments become ingrained in the subconscious so that no more than a quick glance tells the brain whether all is well, or what is wrong. Then maneuvers, such as are required in battle, are performed with no more than the thought of what is needed to achieve an objective, having chosen an opponent. After three years of hard work and practice, it was my joy to achieve this level of proficiency required to remain alive in battle.

Advanced Training

Advanced training continued with the same regimen. My class was the first to skip intermediate training and go straight to advanced training, flying the North American Harvard, AT-6. This was a phase in which I excelled and thoroughly enjoyed.

Toward the end of advanced training in the AT-6, two

Author piloting the AT-6 #209 during advanced training

instructors unwittingly tested me; one might say I was taking a final exam without knowing it. I figured that out later. I was flying in the front seat with my instructor in the back seat. After a while, we formed up with another AT-6 with the instructor flying it and the student in the back seat, just the reverse of our posture. We started a little game. The instructor in the other machine did a maneuver and for fun, I did the same maneuver. After several aerobatic maneuvers where I mimicked each, I presumed the instructor in the other machine was think-

John F. Wilkinson

ing, "I'll get you this time." He did an eight point roll, a maneuver I had never heard of or seen before. This maneuver is very tricky in that if you don't know what you are doing when your wings are vertical to the earth you can 'fall out of the sky' since there is no lift from the wings and one is essentially at full throttle hanging on the propeller. I watched him carefully and then did an identical perfect eight point roll. I am inclined to think that this was a contributing factor to my achieving the award as "Best Flying Cadet" for my class.

During advanced flight training when practicing dog fighting in the AT-6, my opponent and I drifted over a cloud deck. Upon descending below the clouds low on fuel, there was nothing in sight but the swamps of the southern Florida Everglades in all directions. My partner in the other aircraft said to me on the radio, "Do you know where we are?" and I replied, "No"! Whereupon he said, "You choose a direction." I picked a compass heading that just 'felt' right. After flying for some time, we flew right over the center of the airfield. He asked me later, "How on earth did you do that?" Of course, I could not really explain it. I am confident the Lord blessed me with a measure of homing instinct.

Return to England

At the completion of training on 18 February 1943 I was commissioned a Flying Officer, received my wings as a fighter pilot and was presented with one of my most treasured awards - "Best Flying Cadet." Only about 30% of us were commissioned, the rest became sergeant pilots. Immediately thereafter, following another stay in Moncton, I returned to England for operational training.

The return train trip to Canada was in the finest plush carriages with sleeping accommodations for officers. Presumably, whoever was responsible for our earlier train did not want that to be reported to the British.

Back at Moncton in the bitterly cold February of 1943 we were experiencing how frigid it can get in Canada. We had not been issued winter coats or other winter clothing. The dining facilities were at least a couple of blocks from the barracks. It doesn't take much imagination to realize the speed at which we ran to get our meals.

From Moncton by train again, we arrived at New York City on 10 March 1943 to board the Queen Elizabeth luxury liner. There was no need for an escort this time since the QE was at that time faster than anything else afloat. It was equipped with sonar detection and all the latest technologies. In the berth next to the QE the French liner, Normandie was lying on its side. I later learned that while being refitted for wartime use she caught on fire and rolled on to her side. There were thousands of us aboard the QE. In a room designed for two, we squeezed in 22 men. Bunks were built from floor to ceiling so close to each other that one had to slide in sideways. One could not read anything because the bunk above was so close. The aisles were so narrow the only way to pass was for one party to slide into a bunk. There were so many of us we only had two meals a day. Nevertheless, those meals were very good with so much to eat we really did not need any more. We arrived in Glasgow 16 March 1943. Those of us who were pilots went to Harrogate in Yorkshire to await further posting.

Further training on British aircraft and learning to fly over the more complex English landscape was required and I was assigned to Peterborough to fly Miles Master I's and II's. The I's had an inline Kestrel engine; the II's had a more powerful radial Pegasus engine.

One day while sitting on the flight line waiting to fly I was watching a Miles Master II with the Pegasus engine approaching for a landing. The student was in the front seat flying the aircraft with an instructor in the back seat. The student made a rather heavy landing and was instructed to perform a touch and go to try it again. The student pilot applied full throttle immediately to go around for another try. Then, before my unbelieving eyes, as they reached about 50 feet, the engine fell off! Fortunately, they were so near the stall that the aircraft pancaked flat without turning over. Both student and instructor were unhurt, although a little shaken.

The first time I flew the Miles Master II with the radial engine at night I took off with the instructor in the back seat. There are small vanes or flaps around the back of the engine that must be fully open on takeoff for cooling allowing the pilot to see the cylinders in the engine. It was a very dark night, WWII blackout requirements were in force. I calmly said to the instructor that we apparently have an engine fire (those things didn't bother me). He calmly replied that the cylinders get so hot the red glow can clearly be seen at night.

Flt Lt. John F. Wilkinson
1943

From Peterborough, I was assigned to a satellite field a few miles away. I was already recognized as a highly proficient pilot, so when the Miles Master I and II aircraft became unserviceable for students to fly I was given the unserviceable aircraft to fly back to Peterborough for service and repair, which I did without any undue problems. Near the remote airfield there was a very nice country pub which some of our pilots liked to frequent. Some distance beyond, there was a U.S. airbase. Some of their pilots occasionally visited the same pub. One day while pilots from our squadron were enjoying a quiet drink, Clark Gable, the well-known actor and pilot arrived with friends. On a lark, our pilots decided to take Clark Gable's cap. That created such an uproar it nearly became an international incident. Later, on a dark night, those same pilots drove by the pub, tossed the cap into the entrance and that was the end of the incident.

Early Spitfire Training

In early summer of 1943, I, along with other new pilots, was posted to a remote outpost near a small fishing village in Northumberland north of Newcastle. It was nothing more than a collection of Nissen huts (called Quonset huts here in the U.S.) used for training purposes. I was quartered in a hut assigned to commissioned officers. The hut next to us was assigned to sergeant pilots. When we heard the fishing boats returning to the village with their daily catch of lobster, we would walk perhaps a mile down to the village ostensibly to talk with the fishermen and their families. The wives would boil the fresh lobsters that we would then take back to our huts. There we would sit on our bunks, dismembering our freshly cooked lobsters with our "commando" knives then gorging ourselves.

The sergeants in the next hut were quite a rowdy lot, sometimes very noisy. They placed a target on the inside of the double doors and practiced throwing their "commando" knives at it, albeit not very accurately. The result was the door was totally destroyed, chopped up and fed into the central stove. Realizing that the door would be missed, in the dead of night they proceeded to remove the inside door of one of the huts used as a classroom. However, the officers teaching the classes soon noticed the odd disappearance of one of their doors. It didn't take much detective work to find the cause and then the sergeants were in very deep trouble!

Now it was time to learn to fly Spitfires and Hurricanes, both thrilling aircraft to fly, and it was time for some real combat training. Near the fishing village of Eshott was where I had my first encounter with Spitfires. They were battle-weary Spitfire I's and II's. We had already learned the aircraft inside and out, engines, airframes, aerodynamics, machine guns, cannons and so forth. We were taught to reach every control blindfolded so one need never take one's eyes off the battle or in an emergency to reach any of the controls. Perhaps my biggest thrill in flying was the first time I flew a Spitfire. It was not only the first time I flew a Spitfire but also the first time I had flown a plane that I had not first flown with an instructor. This was the crowning task of "Do-It-Yourself" explorations. It gave new meaning to "On-the-job training!" The Spitfire was an absolute joy to fly.

In all I have flown 24 types of single engine aircraft, including three German aircraft, the latter after the war of course. The Spitfire was a single seat fighter that fit me like a glove and it became an extension of me, seeming to respond just to my thought without need of a designed system to control it.

These battle-weary Spitfires frequently had mechanical problems. As in driving old cars, flying weary air machines sometimes led to challenging situations. It is curious how often when one problem flares up it evolves into a series of problems. One day flying up over northern England I smelled smoke, even with my mask on. Not good! I saw a wisp of smoke coming from the gun sight and immediately disconnected it. That took care of the smoke problem. This caused loss of the radio and all electrical systems. On heading back to base, I discovered the pneumatic and hydraulic systems had also failed. It seems that when one thing goes wrong a cascade of others almost inevitably follow. To make matters worse a dense fog from the North Sea had suddenly moved in over the base. I couldn't see anything forward, only what was directly beneath me. With a quick jerk of the stick, I managed to force the wheels down and heard them lock into place. Flying over the airfield, I saw that the short runway was the one in use but I could only see the ground directly below me, nothing forward. I circled repeatedly, each time getting a little lower, memorizing every tree, hedge, even the cows and other possible obstructions. When I was secure enough to attempt this landing, I skimmed over the hedge, touched down just on the end of the runway and when rolling straight, switched off the engine. I sat there in the cockpit, with no brakes, rolling through the fog, hoping to stop before the end of the runway. I did. There I found the crash crew and the

Squadron Commander waiting for me not knowing what had happened. Fortunately, I had a plan and was quite confident in following that plan, so all ended well and safely. Thanks to the Lord, I saved the Spitfire as well as my own life.

When flying Spitfires from Eshott I once happened across a castle that was set in the woods on a remote hillside. To me it looked like a fairy castle it was so beautifully constructed and in such a lovely setting. Every day I would take a detour to briefly circle and take a look at it. I have no idea now where that castle was and never confided in anyone about the pleasure I gained from my little detours.

Developing aerobatics skills and instrument flying are critical requirements in piloting fighter aircraft. Too many pilots when overwhelmed by the enemy have gone into a cloud, only to come spinning out the bottom because they were inadequately prepared to fly on primary instruments. There is a range of instruments available to the pilot to determine important indicators of flight conditions such as how well the engine is functioning, oxygen flow, pneumatic pressure, electrical systems condition, attitude, altitude, airspeed and other parameters. Some of these instruments derive their information from gyroscopes that provide stability to maintain a reference point. The most important of these are the Attitude Indicator and the Directional Gyro (heading indicator). At high altitudes while doing complex aerobatics it is possible that the gyroscope driven instruments can do a thing we call "topple." When gyros "topple" these instruments become useless and reliance on them can be fatal. Proficiency in the use of the primary instruments of altimeter, turn and bank indicator, airspeed and vertical airspeed is critical to survival while attempting to maintain controlled flight in the clouds, and often the clouds are a great place to hide. Although I practiced, I never had occasion in combat to hide in one.

To hone my proficiency at flying on primary instruments, I would fly into a big cloud and perform some aerobatics, then gain complete control using only primary instruments. On one such occasion before correcting for normal flight I popped out of the cloud unexpectedly turning from instruments to visual flight, I looked down for the ground. It was nowhere to be seen. It was above me! There was enough positive G to make me think I was right side up, when I was actually upside down. I really had a good laugh and went back into the cloud. When in the clouds all normal human sensing regarding attitude of flight must be ignored, to pay attention to them leads to sure disaster. Trust in your instruments!

A fighter pilot must retract his wheels immediately on leaving the runway in order to more quickly gain speed and be ready to fight. One day the press joined the Squadron Commander in flying control to watch the new pilots taking off and landing. Being new pilots, most did not get their wheels up until well beyond the boundary of the airfield. The Squadron Commander was furious. He decided to show the press how it should be done. He jumped into a Spitfire and at full throttle raced down the runway. To his chagrin, he lifted his wheels too soon, before reaching flying speed, and settled down on the runway with a very red face! A major repair task for one Spitfire was near at hand.

John F. Wilkinson

While at Eshott, nasty weather from the North Sea set in preventing flying for a few days. Frustration and boredom set in. The Squadron Commander decided to do something about it. He provided us with shovels and pick-axes and had us dig a long trench along one side of the airfield. After we completed it, he then tasked a bulldozer to come and fill it in!

Dogfighting

The next step in my training was in Grangemouth, near Edinburgh, Scotland where we flew Hurricanes and

Hawker Hurricane – combat training at Grangemouth

Spitfires. Both of these machines had acquitted themselves well in the Battle of Britain. It was now time to practice dog fighting and other combat maneuvers, something I enjoyed and entered into whole-heartedly. According to "War Slang" by Paul Dickson, "the term dogfight has been used for centuries to describe a melee; a fierce, fast-paced battle between two or more opponents. The term gained popularity during World War II, although its origin in air combat can be traced to the latter years of World War I." We practiced this frequently.

Two of us would go up, one in a Spitfire and one in a Hurricane and dog fight. Then come down and switch planes before going up to dogfight again. I soon found that it did not matter which plane I was in I always won. It was then when I developed my personal method of fighting. In dogfighting, my way was to push my machine and myself to our combined extreme limits with no restriction on any maneuver. Dogfighting to me, meant climbing, rolling, turning, diving, flying inverted or in any other attitude to achieve an advantageous position behind my opponent to shoot him down. These exercises were the pinnacle of my preparation for one-on-one mortal combat. They magnified the seriousness of close-in aerial fighting where we were to engage the enemy at point blank range; as close as 100 to 250 yards, knowing that one of us was about to die. That is quite different from today where air-to-air missile engagement is from stand-off ranges. When it came to the real thing, God blessed me with the ability to finish them off very quickly.

One time while I was dogfighting in a Hurricane I pulled a high G maneuver, the force of which stripped the fabric off the side of the fuselage from the cockpit to the tail. No problem, it was just a bit drafty so I returned to the airfield. The drag from the missing fabric caused some difficulty in lining up with the runway, so I went around again and this time was able to compensate for it to land safely. I walked over to the control tower to report the mishap. The controller said he had seen my problem but did not want to say anything in case it should worry me! As if I didn't know a good part of my machine was missing!

While flying out of Grangemouth one day I was cruising north over the hills of Scotland. I was quite low and at the top of one hill, I saw a flock of sheep and a lone shepherd. As I circled around the hill, the shepherd waved at me and I waved back to him. For a brief moment, there was an unseen connection between us.

My damaged Huricane from high-G maneuvers while practicing dogfighting

On another occasion, I gained great pleasure flying a Spitfire just above the water for the length of Loch Lomond enjoying the beautiful scenery. Such memories have long remained with me.

Yet another time I was assigned to ferry a Spitfire V to a large airfield in south central England. All went well until nearly there. The radio was not functioning but that was no problem for the moment. Then white smoke began to pour from the exhaust stubs. That is an indication of an internal glycol leak, a very dangerous situation that can lead to fire or an engine explosion. I followed all standard procedures and trailed an American Liberator bomber on final approach to make a landing. Then to my horror the American pilot, disobeying all regulations, instead of turning on to the taxiway he did a 180-degree turn to go back up the runway in the face of my attempt to make an emergency landing. I had no choice but to risk climbing up to circle for another attempt to land. I flew at minimum engine speed just above stalling hoping to keep my Merlin engine running. As I did so, there was an Oxford twin engine passenger plane on final approach. I performed a typical fighter short approach landing and cut off the Oxford who was forced to go round again. After I landed, I went to flying control waiting for the pilot of the Oxford so I could apologize and explain why I cut him off. The Oxford taxied up to flying control and I went to meet him. When the cabin door opened to my extreme surprise, it was a plane full of the highest-ranking Air Marshals! However, they fully understood my emergency and were very generous in their comments. As for the American pilot, he never dared show his face and even neglected to apologize for putting my life in jeopardy.

When a Spitfire XII repairs were completed it had to be flight-tested. Part of the test required clearing the area to assure there were no other aircraft in the immediate vicinity. It was necessary to dive at high speed while watching the ailerons flex upward. There was a red line on the side of each aileron beyond which it should not flex. On this occasion a pilot (I think his name was Fischer) cleared the area, or so he thought, and proceeded to dive at high speed. To his alarm he saw another Spitfire climbing under him with which he was about to collide. Being a strong man he pulled back on the stick with such great force one of his wings broke off. The resulting force broke his straps, ejected him through the canopy, up and out of the aircraft. Fortunately he still had his parachute on, which he opened and landed safely in the back garden of a house. He staggered, bloody and shaken, up to the back door and knocked. The door was opened by the lady of the house. It so happened he had been trying to rent a room from the lady of this house, who would not rent to pilots. It seems she decided that since he went to so much trouble she would take him in as a renter! Do you believe in coincidence or would you rather acknowledge the Lord at work! Think about it.

There was one instance when a new and inexperienced pilot who had recently joined the squadron, somehow got into a flat spin over the airfield. He did not know how to get out of it and as a result he crashed in an adjacent field and was killed. I never saw such a bright fierce fire. You couldn't look directly at it. That was because there was a lot of magnesium in the metal that burned intensely once it was ignited.

A Gift from God

My intent here has been to set down certain facts, rather than to boast of my own capability as a pilot. I recognize that all that I am and all that I have are gifts from God and for that I will be eternally thankful while giving Him all Praise and Honor. You should know that it was not until advanced age that I realized how unusual my flying capabilities were, and that I could do things in flight that most other pilots at that time were not able to do, hence my ability to solo in 8 hours. With all the training; physical, mental and psychological, and conditioning to react instead of think, it is still unpredictable how one will perform and handle actual combat situations. Only God knows and it is He who guides us through the valley of death.

I had tremendous drive and motivation to become a pilot, and more specifically a fighter pilot. When I first volunteered for the Royal Air Force I was questioned by three high-ranking officers. While waiting for my interview I was given a few clues by the volunteer ahead of me when he came out from his interview. It would seem that I gave them all the right answers, but it didn't hurt that I had been schooled at Christ's Hospital. I discovered later that the officers had recommended me as officer material. I was called up a few weeks after the interview.

In the course of my five years in the Royal Air Force training and serving as a fighter pilot there were of course a number of memorable moments outside of the fighting and violence endemic to war. I found a great joy in flying which was greatly enhanced by the graceful and powerful Spitfire. As the birds it was perhaps the one machine in that period that was most incredibly attuned to flight.

John F. Wilkinson

Part III: 41 Squadron

I COMPLETED MY ADVANCED operational training as a fighter pilot in mid-1944 and was posted to Cranfield near Bedford. There I was flying Spitfire Mark IX's, XII's and XIV's. My choice, of course, was the most powerful Spitfire Mark XIV, a wonderful flying machine. I was at Cranfield for three weeks on 30 minutes standby in readiness to leave for a squadron with the Mark of Spitfire they were using. When the call came on 21 July I was gone within half an hour in a Spitfire Mark XII on my way to 41 Squadron at Lympne on the south coast. A transport plane followed with my kit. I was rather disappointed that it was a squadron of Spitfire XII's. Nevertheless, as ordered, I took off in a Spitfire XII and flew to join the squadron.

John F. Wilkinson

Joining 41 Squadron

Lympne was a small grass field on the hills overlooking Romney marshes, farmland and the English Channel. On arriving, I was taken to the officer's mess, a huge mansion on the hillside overlooking the sea. The mansion had been commandeered from Sir Phillip Sassoon. There was no doubt that the mansion belonged to him. Sir Phillip's initials (PS), set in waist high concrete probably at least a foot and a half thick, were all over the grounds and some were even worked into the decor inside the mansion. On the extensive grounds in some places, there were eight to ten feet high hedges six to eight feet across. There were narrow entrances, hidden by large bushes, to passage ways inside the hedges which led to small circular areas for private trysts.

Inside the house there was a library. One of the books, as I recall, was titled "The Life of John Barrick." When the back of this book was pressed a lever shot out from under the shelf. By continuing to press the back of the book while pulling on the lever, a whole section of the ceiling high book case would swing open to reveal a secret stairway. The stairway led up to an open courtyard with a fountain in the middle and small rooms off the courtyard for questionable trysts. Each room had a window designed so that nothing could be seen from outside the building.

Upon arriving at the mansion, I was taken to the bar where some of our pilots were enjoying a drink. One of them, a Belgian, Flight Lieutenant Maurice Balasse, turned to me and waving the stump of a missing leg said, "Just stopped for a quick one on my way to the hospital!" Some months later he was badly wounded again when a 40 mm shell exploded in the side of his cockpit. He returned at full throttle, accompanied by Flt Lt Danny Reid who provided radio assistance and landed just before he lost consciousness. He recovered from that incident also. He was killed the morning of 23 January 1945 while on armed reconnaissance to Münster, Germany when an Fw190 intercepted his Spitfire XIV. Source: Griffon Spitfire Aces, by Andrew Thomas.

While I was with the squadron at Lympne, the V1 flying bombs (often called "doodlebugs") started flying over. We had fighters over the channel shooting them down. Those that got through ran the gauntlet of costal anti-aircraft guns. Inland there were more fighters. The next layer of defense was the barrage balloons over London with their dangling steel cables. At the back of our mansion there was a patio overlooking the marshland below with a stone wall about thigh high. From the patio we could watch the coastal guns in action. The tracers and exploding shells made quite a show. When they did hit a bomb it would dive down and explode in the marshy area below us. You could see the blast wave traveling across the fields and we would duck down behind the wall just before the blast reached us. It was a sport that relieved the off duty time.

Shooting Down a V1

On 11 August 1944 I got my chance to join the squadron in trying to shoot down a V1. The Spitfire XII was not fast enough to catch the 400+ mph V1's (code named "divers") in straight and level flight so we were assigned a patrol line high off the south coast of England at an altitude that would enable us to gain enough speed in a dive to catch one of them. Radar controlled our activities. When radar directed me to a diver, as soon as I spotted it I released the radar and dived at it full throttle. The best way to stop a V1 is to get your wing tip under the wing tip of the V1 and tip it up, thereby toppling the gyros that controlled it causing it to dive out of control before reaching populated areas. On average, a diving Spitfire XII was not fast enough to accomplish this. In this instance I could

German V1 flying bomb

not quite reach its wing so had to wait until it pulled ahead of me. I remained at full throttle and slid in behind it as soon as I could and started shooting right away. The recoil from my guns immediately killed my excess speed relative to the V1's speed. I was dangerously close but determined to stop it. Apparently, I hit the controlling mechanism; it rolled over and exploded in a field. I was so determined to get it I was really closer than I should have been and pressed my gun button so hard, as if it would increase my fire power, that my thumb was sore. I felt a wave of satisfaction that I had saved some potential victims from its intended death and destruction. When the film from my gun camera was developed it showed the V1 as being so large that others were surprised I was so close to it. The film was quite interesting. Once again the hand of the Lord was on me saving me from almost instant death had the V1's bomb exploded. Although this kill was recorded as "shared" with a pilot from 315 Squadron who happened to be in the area, the other aircraft, if he fired on it, failed to get a single hit. When I pulled alongside the V1 it was in pristine condition with not one hole in it. I shot it down singlehandedly and have always considered it my first full victory.

High Altitude Bomber Escort

In September of 1944 our 41 Squadron was finally equipped with my beloved Spitfire XIV's and we were assigned to high altitude Lancaster bomber escort. The long distance flights to the Ruhr district and other targets in Germany with a 90-gallon belly tanks were exciting

41 Squadron Pilots at Lympne September 1944. Author standing fifth from left

events recurring often. We flew top cover for the bombers. We were flying anywhere up to 35,000 feet with no pressurized cabin and no heat, strapped in tight for three hours, sitting on a collapsed rubber dinghy with the consistency of concrete, and under it a parachute. It was hard to relate that to the comforts of home. Our oxygen was not a forced system but rather used a regulator that provided oxygen on demand in accord with the altitude. In summer, flying over 30,000 feet was all right but in winter at temperatures of minus 100 degrees Fahrenheit or lower it was very uncomfortable to say the least. We had electrically heated slippers in our flying boots, waistcoats and gloves. I wore four pairs of gloves, first chamois for easy removal in case of fire, then the electrically heated gloves, then my usual deerskin gloves and on top of them big standard issue elbow high leather gauntlets. I wore silk stockings (as many did), long johns, a lined leather waistcoat, wool battle dress and ankle to hip heavy wool leggings (knitted by my grandmother of heavily treated seamen's wool). There wasn't room in the cockpit for the heavy wool lined jackets worn by the bomber crews. We only wore those on the ground. Then, of course, we wore the Mae West life preservers when flying over water, which meant every flight from England to the Continent and sometimes even when just in proximity to the sea. In the pouch behind my head in the Mae West I carried a tin of water - necessary if downed over land or water.

In the tunic of my battle dress there were two inside pockets, one on each side. In each one of those I carried a survival pack. The packs were curved plastic boxes shaped to fit against the chest. Each pack contained a week's supply of high energy, highly compressed food, water purification pills and bag, fishing line and hook, sewing needle and thread (In those days trousers were held up with suspenders and buttons. If the buttons came off one would be rather conspicuous walking about holding up one's trousers!). Two of the less critical buttons could be removed and were designed such that when one was balanced on the spike in the middle of the other they formed a compass with small dots to indicate north and south, vital if one was downed and able to evade capture. In addition, I carried a thin rubberized pouch that contained silk maps and a small hand knife. The maps were printed on silk so they would not make noise if one was caught and patted down.

Our flying boots were wool lined. The shoe portion laced up and looked like an ordinary black walking shoe. The legging part of the boot contained a small knife that could be used to cut the stitching and remove it making one less conspicuous during escape and evasion. I carried a large double-edged knife strapped to my shin with the handle above the legging part of my right boot. This was

All smiles and ready to fly

for easy access should I have to bail out and get caught up in a tree. I could then, with great care, cut myself loose. Along with all this I carried my standard issue revolver side arm and belt and a pistol in a shoulder holster. Then, of course, there was the pilots cloth helmet with built in head phones and detachable oxygen mask with microphone. These got plugged in along with the electrically heated clothing after entering the cockpit. In the undesirable possibility of being downed, whether over enemy territory or rough seas, preparation can be the key to survival. I am thankful I never had occasion to use these emergency devices. The hand of the Lord saw to that! Don't take that lightly; think about it.

John F. Wilkinson

We wore goggles with interchangeable clear or darkened lenses, the latter needed since much of our flying was above the clouds and in bright sunlight. The flat lenses were held in a metal frame with padded leather around the edges for comfort when worn for long periods and to absorb sweat. On the edges of the main shatterproof glass, there was a second lens on either side angled at perhaps 20 degrees to allow good peripheral vision so necessary to see the enemy first as well as when engaged in a dogfight. I still have mine. The Lord blessed me with an uncanny ability to see things faster and at greater distances than my contemporaries. I was reminded of this when I read a book called "Blood, Sweat and Valour" by Steve Brew, a definitive history of 41 Squadron from 1941 through 1945. In the book, WO John A. Chalmers was quoted as saying I was "...blessed with marvelous eyesight. Picked up things well ahead of anyone else." I didn't realize it at the time but it helps explain why I could sometimes call the break to face the enemy before the others could see them. This was especially important when the new Messerschmitt jet ME-260's came into service. They were fast, but if they made one turn, whoever was closest got to shoot it down.

Notwithstanding all this gear, we had a grandstand view of the action, the dreadful destruction of German cities and the loss of many bombers. The heavy 88-millimeter guns would shoot at us occasionally. When they came close, the exploding shells made a "crump" noise but a little dodging and a quick change of altitude kept them from getting close enough to do damage. Our presence above the bombers kept the German fighters away, but what the Luftwaffe didn't know was if we had to fight we would not have enough fuel to get home. With a 90-gallon belly tank we could fly for three hours. We were usually at our absolute maximum range.

Our high-powered engines were not intended for such slow flight, minimum cruise, as usually required when escorting bombers. A coating of lead would be deposited on the spark plug electrodes as a result of burning 150 octane fuel at reduced RPM's. To burn the lead off required full throttle for a few seconds every fifteen minutes. This did not always work leaving one with a

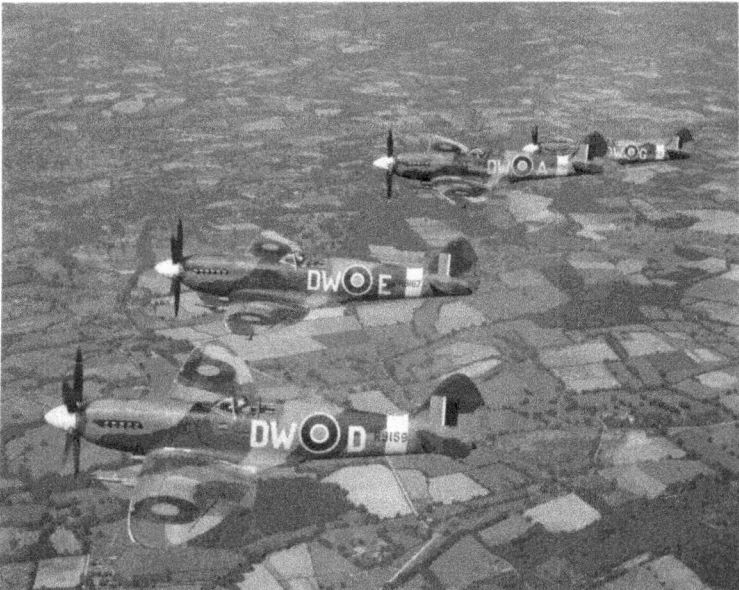

Spitfire XIV's over England

rough running, poorly performing engine. On one occasion when returning across the English Channel, watching the Cliffs of Dover draw nearer with my fuel gauge reading empty, I was ready to bail out. One does not want to ditch a Spitfire because the large air scoops under each wing take it straight to the bottom and the pilot with it. Given that one does not have enough altitude to bail out, one procedure we learned for avoiding visiting Davey Jones' Locker was for the pilot to open the cockpit hood and stand up just before hitting the water. The force of the near instant stop of forward motion would pitch him out in front of the sinking plane. In this instance the fuel gauges were reading zero. After refueling the ground crew determined that I had less than one minute of flying time left.

On another high altitude trip to the Ruhr my plugs leaded up so badly I was burning fuel so fast that I didn't think I would make it back across the channel so I landed at an alternate designated RAF airfield in France. On landing I sent a message reporting my location. However, it turned out that the message didn't get through and I was reported as missing in action. Apparently that was when those in charge found out that my sister Joan, a Radar Officer, was stationed at the same base although not in the same area, which resulted in her being posted to another base! At the airfield in France I met several other pilots I knew. There was a bountiful supply of liberated French Champagne flowing like water, it led to a happy evening and a rather over indulgence of the bubbly. So the next morning, not accustomed to alcohol use and rather the worse for wear, I supervised the repair of my engine and returned home.

One day we took off with the squadron at full strength, planning to rendezvous as usual with the bombers over the French coast. However, my Australian friend Flight Lieutenant Danny Reid and I both developed engine trouble over the channel and had to set down at

Manston, a huge aerodrome on the southeast coast of England designed for returning bombers in trouble. Our Spitfires could not be repaired until the following day, so Danny and I decided to go up to my home in London. Believe what you may, but we did not plan this. When we phoned, we learned there just happened to be a party on at home that night which we attended. We were in our battle dress and flying boots, etc., but we did leave our guns (which we were rarely without), knives and Mae-Wests behind. That was when Danny met and became friends with my sister Eleanor. We had a lovely evening and returned the next morning to await completion of the repairs to our Spitfires. Danny dated Eleanor for a while and although quite smitten they did not marry, but that's another story.

When over enemy territory it was necessary to fly in open battle formation. To change course, this required crossover turns. It was quite a tricky maneuver when flying so far apart in battle formation, but the formation permitted us to watch for the enemy and cover each other from surprise attack. A squadron of twelve Spitfires might spread up to a couple of miles across the sky. To make a 90° left turn for example, the aircraft on the extreme right would turn first, followed by each one and so on allowing the same space to be retained between each of us. When the maneuver was completed we would be spread across the sky in the reverse order.

On one occasion we were required, as top cover, to escort Mosquito twin engine path-finders to Walcheren Island in the mouth of the Scheldt River which emptied into the North Sea. The island was a hot bed of German anti-aircraft guns that often caught our nighttime bombers heading for German cities. The Mosquitoes dropped their flares leaving towering trails of smoke to guide the bombers, and then left. We had to hang around while the bombers arrived and dropped load after load of bombs breaking the dykes and flooding Walcheren Island. This literally sank the island and no further action was needed to eliminate the anti-aircraft guns. It was quite a show and interesting to watch!

At the airfield we had a Nissen hut where we gathered while waiting to fly. In that hut we had a stove that we kept burning when the weather was cold. One day the fire went out and my flight commander, Flt Lt Ross Harding, picked up a bottle containing a mix of oil and aircraft petrol that we used to pour on the cold coals then stand back and toss in a lighted match. He poured some into the stove but as it happened the stove was not quite out. The flames leapt up and set the bottle on fire so that the flames were coming out of the neck. Well, that wasn't a problem. He just went to the door and threw it out. I was sitting by the door and it is the only time I ever saw the blood literarily drain from a man's face and turn pure white. I got up, looked out the door to see what could have caused this phenomenon. A petrol bowser (large tractor drawn tanker) had been parked right outside and the flaming bottle had landed right under it. I grabbed a fire extinguisher and quickly put it out before anything worse happened! Why the bowser driver chose to park it there escapes me. Once again the hand of the Lord prevented the loss of all of us in that Nissen hut.

While in Flight Training at Riddle McKay Aero College I became good friends with a fellow student Alfred George "Bunny" Henriquez (later Flt Lt) from Jamaica and some of his family who lived there in Clewiston. When we left Florida and returned to England, Bunny was trained to fly bombers and I transitioned to the Spitfire fighter.

After I joined 41 Squadron on the southern English coast, as I mentioned earlier, our task consisted mostly of escorting bombers. During this time I developed an urge to visit Bunny who was stationed toward the north of England with 630 Squadron. There was a repaired Spitfire IX that needed testing and I got permission to test it by flying up to see Bunny. I almost didn't make it! When I took off the left wing was so heavy I had my shoulder against the side of the cockpit and both hands on the stick, using all my strength to maintain level flight. I circled and landed and a mechanic came out to see what was wrong. When I told him he told me to stay in the cockpit. He got a hammer and block of wood and proceed to bang mightily on the trailing edge of the wing. I took off again and it was now in perfect trim.

I had a good visit with Bunny and inspected his Lancaster. When I took off to return to base he was standing by his Lancaster bomber. I circled to head south and as I looked down on him standing by his Lancaster I had a strong premonition that he would soon be killed. A few months later word got to me that he had been shot down and killed. His family wrote and told me his small daughter had awakened one night screaming to pray for daddy. It occurred the same night he was shot down. As fighter pilots, our chances of surviving were considerably less than a man running across a heavily traveled street dodging speeding cars. I decided that from then on I would not form any more close friendships with my flying companions. Emotions, especially anger, can cloud the objectivity and cool response needed to meet the requirements of life and death combat.

On one mission our whole squadron took off to rendezvous with bombers across the English Channel. There was a heavy overcast so we flew in close formation to climb up through it. I was number two to Squadron Leader Benham. When we climbed up into the clouds it was so dense we could hardly see our wingtips. I closed in very tight and eventually we came out into the bright sunlight. Looking around the Squadron Commander and I were the only two together. All the others had broken off and came popping up from the cloud at varying distances from us to reform again.

Both Flt Lt Ross Harding and I had engine trouble one day when leaving on a mission to cover bombers over Germany. We landed at Manston airfield on the east coast of Kent. Manston had a huge, long, wide runway so returning damaged bombers could crash land there. After our Spitfires were repaired, we flew back to Lympne, our home airfield. Ross, not knowing that I was very proficient at close formation flying, told me to tuck it in tight, which is exactly what I did and thought nothing of it. My wing tip was just inside his wing tip, requiring considerable skill due to the airflow coming off his wing. When we got down he said he was scared stiff and didn't dare move a muscle! I held back a chuckle.

Part IV: Aerial Combat

After the British Army pushed into Belgium in December 1944 and after long hours flying top cover for the bombers to the Ruhr and other targets, 41 Squadron moved from England to Belgium 4 December 1944 just before I went to required air crew training. We first flew in to Evere, and the next day we hopped over to Diest - Schaffen, a few miles away. The field was so small that a yellow square was placed at the end of the runway where we had to land. The yellow square was exactly the size of our Spitfires. Coming in over a tall stand of poplars we were required to land with all three wheels squarely and firmly on the yellow target or we would run out of runway and find ourselves in a hedge with a ditch and road beyond it. That's no problem for a proficient pilot, but I must say a few times careless pilots did not have all three wheels placed firmly on the yellow square. The results were obviously predictable.

From this small grass field at Diest, Belgium we began attacking all manner of ground targets behind enemy lines. While at Diest a call once came to scramble four aircraft quickly. We were ready to respond! We ran and climbed into our Spitfires. There wasn't time to get to the correct end of the runway so we took off downhill and downwind toward the poplars. It was the first time I increased the throttle to full power on takeoff that quickly, a very dangerous maneuver since the torque was so great it could flip the aircraft. We all made it up and over those tall trees but were unable to find the enemy who apparently had fled – or it was a false alarm!

During our stay in Diest we lived in a monastery. We had the ground floor and the monks had the upper floors. We would frequently rise at 3:00 a.m. to leave for the airfield and be ready to take off at or shortly before dawn. Even at that early time we would hear the monks moving about upstairs for their early duties. Since we had no means of bathing we learned to barter with the nuns down the street for the use of their bathroom once a week. That was good use of chocolate and bars of soap. I had fun shocking people at home telling them I lived in a monastery and was naked in the nuns' convent!

Dinghy Training

Shortly after arriving at Diest, Squadron Leader Benham called me in to tell me I was being sent on a temporary posting to Air Crew Officer's School Hereford close to the Welsh southeast border. I tried to get out of it and complained bitterly to him. Whether this was intended for me specifically or if it was some training to which squadrons were required to send a pilot I did not know. I suppose since I was one of the newer members, or perhaps the newest, the "dirty" deed was given to me.

When I arrived at this temporary posting I found I was one of about 30 pilots on the course. The purpose of the training was lifesaving and dinghy drills for both fighter and bomber pilots. We were issued coveralls as our swimming apparel. Then we trundled over to the local swimming pool. This was December, in the dead of winter, and the pool was not heated! The course was about 10 days long, much of it in the water, wearing only the coveralls. We had to swim the length of the pool under water, grab a pilot who was floating there and tow him back to the shallow end. We had to swim under water to alternately small and bomber size inflated dinghies coming up underneath them and righting them, not an easy task in the deep end. Finally we had to tread water for ten long minutes. They were grueling exercises. The cold was bone chilling. So demanding was the course that only eight of us were able to complete it. In retrospect, the training was quite valuable as it raised our level of confidence in survival at sea – and the North Sea can be quite cold!

One day after we were released from coursework, a few of us walked down the street and entered a cozy British pub. A pleasant bar maid took our orders. She must have noticed that I was I blue and shivering as I asked for something that could warm me. She poured me some ginger wine and very remarkably not only was it hot in my stomach but the warmth incredibly started to spread to my extremities coupled with a wonderful stirring of the blood flow. The eight of us were given certificates and we went off on leave. I, for one, enjoyed Christmas at home in London on extended leave.

When I returned to the squadron on 17 January 1945, the squadron had moved to a better airfield near Ophoven, Belgium. On my return, I learned that on New Year's Day, just a few weeks earlier, the Germans mounted a massive attack on airfields. They told me a captured Spitfire led the attack on our airfield. Interestingly the Germans thought there would be enough New Year's Eve carousing to keep the pilots on the ground on New Year's Day. They miscalculated in our case. Fortunately, the full squadron was already in the air deep into Germany so only the remaining backup Spitfires were destroyed. It did, of course, take time to clean up and get more replacement Spitfires sent over from England, hence the extension of my leave.

Ground Targets

Taking off from Ophoven on a pleasant but cold afternoon on 22 January 1945, four of us flew our Spitfire XIV's deep into enemy territory where we attacked targets of opportunity, fulfilling our official squadron motto "Seek and Destroy." We mostly flew in groups of four or six - sometimes eight. On this occasion I destroyed a few trucks and went on to attack a train, one of our primary ground targets. We were especially determined to disable the engines needed to deliver the supplies of war. The most sensitive and irreplaceable part of the engine was the cylinder driving the wheels. While aiming at the cylinder the spread of our shells would also blow up the main boiler sending up a huge cloud of steam. That was quite satisfying to us pilots. This particular afternoon we came across a freight train and proceeded to attack it. When attacking trains we always attacked and disengaged from different directions to make it difficult for the anti-aircraft gunners to track us. The enemy, knowing our preference to attack trains, would sometimes set up a decoy train running back and forth beside a heavily wooded area. The woods, of course, were loaded with hidden anti-aircraft guns, hoping that we would all attack from the same direction making it easy to pick off the later attackers.

About to close the cockpit of the Spitfire XIV

Picture now, the intrepid fighter pilot in his powerful fighting machine, constantly turning his head and scanning not only the ground below for targets but also the full range of the sky for enemy aircraft. Between wars one and two, much was made of the dashing pilot and his silk scarf. Those were much more functional than indicated. We wore our coarse wool battle dress and the constant turning of the head would cause very sore necks, hence the need for a silk scarf. Initially I would tie my scarf in tight knots to its very end. Even so by the time we got home it was totally untied, hence the need for a scarf pin.

Sitting strapped into the cockpit, the pilot finds the instrument panel in front of him and above that, the bulletproof windshield with the lighted gun sight superimposed on it. When acquiring and selecting a target on the ground the pilot wheels his machine over to point the entire airplane at the target. Then, as he gains speed hurtling toward it, he places the target in the center of his gun sight. At the right moment he opens with perhaps two seconds of fire with a destructive burst of cannon and machine guns before pulling hard away to avoid possible flying debris or anti-aircraft fire. He then climbs up to his cruising altitude and proceeds to the next target of opportunity.

Spitfire XIV cockpit

We were flying out of Ophoven when we were ordered to attack enemy supply lines in a specific sector. The target area was directly behind the Battle of the Bulge in the Ardennes forest. The weather was terrible, the anti-aircraft fire murderous and it was hard to see the targets, but we did what damage we could before returning to our base.

One day, diving down toward a railway engine, while scanning the area for anti-aircraft guns and watching for enemy fighters, I released the gun safety catch and switched on the gun camera to record the action as the guns are fired. I looked through the gyroscopic gun sight projected onto the bulletproof windshield, over the length of the powerful engine, through the propeller spinning so fast it is no obstruction to the view. As my view of the locomotive grew larger, I maneuvered to place the center dot of my gun sight on the cylinder that drives the wheels. At the precise moment to engage, I pressed the firing button and felt the powerful rumble as the shells and armor piercing bullets sped on their destructive path. Having destroyed the engine, I then raked the freight cars with can-cannon and machine gun fire to destroy as much of the freight as possible.

This time my aircraft was struck by exploding projectiles from the anti-aircraft fire that damaged the huge radiator under my port wing. However, the hand of the Lord was once again upon me as amazingly the radiator was not punctured which would have shortly brought me down with disastrous results.

On this particular day, having attacked a train and other ground targets, I spotted a large covered truck traveling a country road. Alerting the others, I dived to the attack. While I damaged the truck, I was not satisfied with the results and circled around for another attack. As I approached, I saw some men running for the ditch dragging another man with them. Ignoring the men, I opened fire once again on the truck and there was an enormous explosion, such as I had never seen. I quickly pulled up to avoid the blast and debris I then returned to cruising altitude. It would have been an immoral waste of life to attack the men in the ditch who survived and whom I presumed were probably lowly conscripted privates. From my camera gun film and my report, it was later determined that the truck was carrying a load of land mines to the front. It was very satisfying to know that those deadly explosives would not be used against our soldiers.

War is a deadly process and is never to be taken lightly. Facing the stresses and dangers of battle weighs heavily upon those faced with daily combat. Even television cannot convey the deadly nature of real war. However, there is the rare occasion when one finds a little relief in a lighter moment or an absurd situation, even while engaged in an attack.

While attacking other targets, three of my guns jammed one cannon and both 50-caliber machine guns, an unusual occurrence. As we headed home, I spotted a small canvas top van driving slowly along an open road with no trees or other cover around. I decided to have a go at it and headed down to attack. With only one cannon on my right wing firing, the powerful recoil kept kicking me off my aim. As I tried to compensate with rudder pedals, my shells wandered and were hitting all around the truck. The driver just kept on going. I made a second pass with the same results. The driver just kept tooling along never changing his speed or direction. On my third pass, I started to laugh at the ridiculous situation and thought to myself "You've really earned it today!", and headed for home. I did not observe a single direct hit on the van. It was certainly his lucky day! He must have thought what a terrible shot I was and I wondered if the driver was deaf and blind. One does not feel any animosity toward these targets, just the need to destroy the enemy's ability to wage war. Such is the futile waste when one must engage in war to prevent the destruction of our way of life and to preserve our freedom.

It was 2 February 1945. It snowed the day before and a new day was dawning clear and very cold. We were operating out of an airfield in Holland called Volkel. Eight young men, each carrying a parachute, walked toward the shadowy forms of the Spitfires with their big five bladed propellers. Standing by each machine was a member of the ground crew, some of whom had worked all night to repair and ready the aircraft for their twice-daily sorties behind enemy lines.

The Spitfire XIV was the most powerful and deadly conventional fighter of the war. Its big 2,050 horsepower engine could only be started with cartridges that would turn the engine over so the pilot could nurture it into life. One after another, each of the eight pilots fired his starter cartridge. However, the intense cold had crystallized the copper safety valves causing six of them to blow out rendering the craft unable to start.

I fired my starter, the engine spun and leapt into life. I saw that one other engine, Squadron Leader Benham's, also sprang into life. The other six were unable to start their engines. S/L Benham saw the situation and said to me on the radio "Let's go anyway." I gave him the thumbs up. Like him, I was eager to get going and followed him as he taxied out to the runway.

In wartime, with the Germans always listening, radio silence was maintained and used only in emergencies, such as to call a break to face attacking aircraft. Once in the air Benham and I immediately went into wide-open, side-by-side, battle formation. In this way, there are no blind spots to permit an enemy aircraft to sneak up on us.

Our mission was to attack any vehicle that moved and any aircraft that flew in order to disrupt enemy supply lines. Flying from 12,000 to 15,000 feet, we were low enough to attack ground targets quickly and high enough to give us time to avoid anti-aircraft fire. We were having quite a field day shooting trucks and trains when I spotted a staff car, a choice target since it meant high-ranking officers capable of the decision-making that could result in casualties to our side. They saw me coming and ran into a field. On my first pass I destroyed the car and on my second pass took care of the officers with 50 caliber machine guns and 20 mm explosive cannon shells, the only time I found it necessary to attack personnel on the ground. As we proceeded in open battle formation, we inadvertently crossed a heavily defended area. Benham passed over a small town, while I was passing over a hidden grass airfield. Instant pandemonium! Anti-aircraft fire all around me!

In such cases the best defense against anti-aircraft fire is to use the full climbing ability of the Spitfire XIV. Heading up at 8,000 feet per minute, one can usually climb out of trouble. However, in this case I could see the tracers and explosive shells gradually catching up with me. I instinctively flipped the Spitfire on its back and pulled a high "G" maneuver to dive just as they scored a couple of hits. I felt the bullets hitting my machine. There is no mistaking the sound of it. Naturally, I didn't immediately know the extent of the damage. With the propeller in fine pitch and maximum power the engine was screaming as I headed vertically straight down with the ground coming up at me over 500 mph. I pulled out of the dive very sharply at ground level and leveled off just inches above the grass (one of the many fine qualities of the Spitfire) thereby avoiding disastrously mushing into the ground. I could see the tracers going over the cockpit hood, but I was so low that if they depressed their guns any further they would be shooting each other. In a split second I was over a hedge and away.

Taking stock, I let Benham know my aircraft had been hit and that I had lost oxygen. He looked me over and determined that fortunately there was no serious damage. However, we played it safe and returned to base. This time one bullet passed through my left wing and another passed just an inch behind my cockpit seat and cut the oxygen line. It was just another incident in a day's work. Much more than instinct though, it was the protecting hand of the Lord guiding me away from further harm. The same protecting hand of the Lord is just as available today to those who turn to Him. Consider it as you approach your daily work.

Much can be said about the futility of war, but when one's homeland is attacked and freedom is in jeopardy, those who can, must spring to arms. In my case I chose to become a Royal Air Force fighter pilot. Those were indeed dangerous and exciting times, not to be wished on anyone and thankful to have survived, thanks to the good Lord. The hand of the Lord was indeed upon me.

Until one experiences combat, no matter how well trained, it is utterly unpredictable how one will react. Being shot at fine tunes the senses beyond description. The sounds, smell and feel of combat are absolutely unparalleled by any other endeavor on earth. Only those who have experienced it first-hand can truly appreciate and understand the power of adrenaline as it courses through the body raising every nerve to the peak of efficiency.

After initial engagements with the enemy it becomes clearer what must be faced. However, this does not lessen the stresses brought on by these life and death occasions. One never knows from day-to-day, while shooting and sometimes being shot at by anti-aircraft guns; attacking ground targets, flying through flak and dogfighting with enemy fighters, whether one will live to see another day, or worse, to be horribly maimed. Self-confidence and courage are essential to survival.

One can only depend on one's skill and the mercy of the Lord. One thing the experience of combat does is that it makes one truly aware of the beauty of God's creation. One no longer takes it for granted, since one does not know whether there will be another day to see it. In the course of the winter of 1944/45 the squadron experienced rather significant losses to anti-aircraft fire, whether shot down or injured and unable to return. I was one of the fortunate ones who always returned uninjured, although there were some very narrow escapes with heavy damage to my aircraft.

A squadron in the air is normally comprised of twelve single seat fighters. In order to be able to keep the squadron at full strength to fly two or three missions a day, about twenty aircraft and 25 pilots were required. This level of man and machine facilitates the time required for aircraft maintenance and relief for the pilots. However, by early spring of 1945 our losses were such that we had only eight active pilots and six available planes. Each day we flew well behind enemy lines to attack his supply lines and shoot down his aircraft. Each day we brought our Spitfires back, some with bullet holes and some hobbling on the wings of angels, barely able to make it. The ground crews would watch anxiously to see how many of us returned. Then as we landed they looked to see if the patches over our gun ports had been shot away, as they always were, indicating that we had engaged in combat.

Then they would work all night to repair them for the next day's sorties. There were some days when there were not enough serviceable planes to fly. It was at that point that we were pulled out of action to reform, retrain and refit.

As we left the arena of daily combat in March of 1945, although a temporary reprieve, it was as if a great weight had been lifted from one's soul. It was good to know that for a few days at least, one could anticipate living without the possibility of not seeing the end of the next day. However, none of us ever spoke of this. One gains a

Pilots of 41 Squadron March 1945.
Author second from left.

completely different view of life and a new appreciation of God's wonderful world given to us to live in and share, although the sharing has been subverted somewhat. The moral here is: don't wait for possible life threatening circumstances to realize the beauty and wonder of life here on earth and the value of companionship.

John F. Wilkinson

It was not until I had aged a bit that I realized in order to handle my own fears I needed to effectively block out the losses of my fellow pilots. Only once did I fire in anger and that was after my best friend Bunny Henriquez, a bomber pilot, was killed and then shortly after I almost was shot down myself. Otherwise, it was a time of calmly and skillfully handling every situation that confronted me and pushing my Spitfire to its very limits.

Before returning to England from the Netherlands we each made purchases of little luxuries that had been unavailable in England since the start of the war. I bought some bottles of Chanel No. 5 and some fruit. One of our pilots bought bottles of champagne. There is no room in the cockpit of a Spitfire for any baggage, so I put the perfume in the pockets of my Mae West life preserver and placed the bag of fruit on the housing and lever to raise and lower my wheels. The fellow with the champagne put the bottles in the wings with his cannons and machine guns. On the day we returned the weather deteriorated and we had to climb over 20,000 feet to get over the storm. The altitude made the perfume bottles leak and the corks pop out of the champagne bottles. Champagne was dripping from my friend's wings and I smelled worse than a perfume factory. Even as fighter pilots, we had to go through customs. However, the customs officers were very lenient with fighting men. When asked whether there was anything to declare, the answer was "No!", even though the air was thick with perfume. One pilot tried to declare 1,000 cigarettes and the customs officer replied, "Did I hear you say 100? No duty required!"

We were assigned to an airfield near Warmwell in Dorset during March of 1945 where we experimented with dive-bombing using Spitfire XIV's with small practice bombs under each wing. That in itself might be considered quite hair raising! There was a target placed in the sea a few yards off the coast. We would approach the target at about 12,000 feet, roll into a near vertical dive with the target visible above the nose of the aircraft. Then throttle back with propeller in fine pitch to reduce the buildup of speed in the dive. There are no dive brakes on the Spitfire that had very clean lines. As a result, speed in the dive builds up very rapidly. Next, one pulls up slightly until the target is no longer visible then releases the bomb. This was sort of a "by guess and by gosh" system not capable of producing good accuracy. After that, pull out hard to avoid hitting the water and becoming a wet statistic. The pull out is the hairy part. With the seat down as low as it will go and feet on the top steps of the rudder bar, one pulls in one's stomach as hard as possible to keep as much blood as possible to the brain. One pulls back on the stick until one loses one's eyesight but no harder in order not to lose consciousness. When you figure you are on the way up, then ease back on the stick to get your eyesight back and you are home free! Your right hand must grip the stick with a death grip and the left must grip the throttle control. If either comes loose from the controls, the gravitational pull is so great it cannot be lifted again. In addition, the head must be held rigidly vertical. If the head gets off center it will be forced down onto one's chest and cannot be lifted until one eases off from the pull out. The net result was that we literally bent the wings on twelve Spitfires. The top surfaces of the wings were severely rippled. It was a miracle that no one lost a wing and was killed. That was the end of that experiment and

John F. Wilkinson

our "relaxed" time for reforming our squadron. So, with a fresh supply of pilots and Spitfires we returned to this business of fighting a war!

During the next few weeks we moved frequently so our squadron could keep up as the army moved forward into Germany. By moving, we could remain closer to the front lines that gave us a shorter distance to fly to reach the enemy, more time to engage him in combat and ability to save fuel. My logbook shows we made sorties form Eindhoven and Twente in Holland before we moved to Celle not long before the end of the war. Almost all these sorties were engaged in attacking ground targets such as ammunition and mechanized vehicles as well as trains and locomotives.

In the latter weeks of the war 41 Squadron moved to Celle in northwest Germany where the horrendous Bergen-Belsen concentration camp was located. It had just been liberated by the British Army. Belsen was one of the worst death camps. It was an appalling sight never to be forgotten. The barely surviving prisoners were walking skeletons. Everything possible was being done for them and the official photographers were there documenting the horror of it. I saw Belsen with my own eyes, so I am witness to the atrocities of the holocaust. Although some deny it, I know it happened! I was there!

Foul Weather Flying

In England and coastal continental Europe the weather tended to be somewhat capricious and it sometimes reached the point where it was not safe to fly. However in times of war there are occasions when it is necessary to shed the restrictions of safe minimums in order to support the ground gaining forces. Consequently we were quite often called upon to fly in conditions that put our training and courage to the test. Dense fog where visibility was near zero was the worst of conditions. Our ground crews would set out a line of gooseneck flares along the side of the runway. If we could see one flare to start our take off roll we would take off hoping by the time we got back the fog would have lifted enough to see the two or more flares needed to orient oneself for a safe landing – that is, if we could find our way back. In those cases the Lord guided me back to base each and every time. There was one flight of four who were not so fortunate however. They had to bail out and float down through the fog, happily on our side of the lines. Then at other times we were faced with near blinding rain, hail or snow. Again at other times the clouds were so low that descending through them became problematic as it was essential that when one broke through one regained instant proper orientation so as to be clear of obstructions and able to land. While there were a few airfields in England with a primitive audio beam approach system designed mainly for returning bombers it was never available to us on continental Europe.

While stationed in Holland the worst weather always came in off the North Sea. This was typically deluging rain, accompanied by dark clouds down to the deck. During one of these storms we were assigned an emergency mission to support Army elements that were pushing into Germany. According to the report they were being attacked by German fighters. The squadron launched a flight of four, of which I was one, to engage the Luftwaffe aircraft. The ride was incredible. Flying low so we could navigate using landmarks, while fighting heavy rain and turbulent winds, we flew directly into the base of a thunderstorm. The words to adequately describe what it was like escape me. We were being tossed and battered about and then pummeled with heavy hail. It is unlikely that any aircraft other than our Spitfires or similar could have survived. When we arrived at the front lines the visibility was so poor we could barely see. Of course there were no German fighters. Who would be foolish enough to be flying in this kind of weather? After briefly searching the area we set about trying to return safely to our base. We all did, I am glad to say, in spite of our overheating engines.

The hail had done so much damage to our machines that we were in danger of not returning at all! The covering on the propellers had been stripped off exposing the laminated wood. Windshields and canopies were battered. There was damage to wings and tails, but the worst was the damage to our radiators under each wing. The honeycombed metal in the radiators had been completely flattened to become solid metal facings not allowing any cooling to our engines. The cameras in our wings had been shattered and paint stripped from all the leading surfaces. We had four unusable Spitfires. All in all it was a mighty experience in survival. We decided the Army was likely being shelled by artillery. Once again under any natural progression of events we would not have been able to return home. It was indeed by the hand of the Lord we were kept airborne until we returned. Think about it and the alternatives.

The Spitfire XIV in the Air-to-Air Role

Soon after I was promoted to Flt Lt I became engaged in my first actual individual air-to-air combat beyond the V1 experience. On a gloomy overcast morning 16 April 1945, five of us were conducting a sweep around the Lake Schwerin area. Although since moving across the channel we had been destroying ground targets, I did not record that we found any ground targets of opportunity on that day. However, as we flew over enemy territory, we spotted three Focke-Wulf Fw190's, Germany's premier fighter, in battle formation. It was the only aerial combat in which I was involved where we were not heavily outnumbered, yet while I was with 41 Squadron, as far as I can recall, we lost very few Spitfires to enemy fighters. The three Fw's were lower and directly ahead of us, a perfect setup. Clearly, they did not detect us as we closed in on them. I picked the one directly in front of me, closed in on him and opened fire with a burst of

German short nosed Focke-Wulf Fw190 fighter

50 caliber machine guns and 20-millimeter cannons. Little stars of shell bursts appeared on his fuselage and he headed down in a gentle dive streaming oil. It was the first time I had met the enemy in the air. I didn't even have to fight with him. I would have gladly done that. It seemed all too easy. I spent more time looking behind me

to be sure any more in the area did not creep up on me as I had on the one before me. I was very close when I fired. My first shots damaged the engine and possibly killed the pilot outright as he made no attempt to take evasive action. His engine immediately poured out a cloud of engine oil covering my windshield and some washed over my canopy. The oil, of course, had coated the lens of the camera that would have recorded the kill. I followed, determined to finish him off, but since I could no longer see forward I slid to the side so I could see him through the smeared canopy and then slid back behind him to fire again. I followed him down to the ground where his flaming wreckage spread across a field. It was all over very quickly. My four companions shot down the other two Fw's.

Although combat training prepares one for the psychological aspects of killing and the subsequent loss of life, the hardening of feelings does not deter the joy of a victory. I was welcomed back with elation by my armament sergeant who jumped on my wing to congratulate me on my first victory. A sense of well-being really emanated through me from the fact that there was one less aircraft that had the potential to bring harm to my countrymen. At the same time however, I was clearly cognizant of the fact that the Lord had permitted me to take a life in the defense of my country. I recognized this as an awesome responsibility and one that shaped my respect for life. Further, I was now fully indoctrinated to the horrors of combat and in the name of the Lord, had the potential to contribute further to the overall effort of ridding the continent of those who would dictate our freedoms.

We had a healthy respect for the German pilots and we entered into combat with complete confidence of victory because of our superior machines and training. We were very respectful of humanity with no intent of wanton destruction - not true of the Nazis. We would not shoot at defenseless civilians. Our aim was to cripple the Nazis capability to wage war. I sincerely believe that not one of us would ever shoot at a pilot hanging helplessly in his parachute. However, I was not aware that any of the pilots I shot down were able to bail out.

On 20 April 1945, I shot down two short nosed Fw190's northwest of Berlin. Seven of us came across more than eight of them and we attacked immediately. When the battle was joined, I quickly performed a tight steep turn defensive maneuver while I picked out an opponent. I saw one of them in a steep climbing turn with his guns blazing at me. "OK," I thought to myself, "you're it." We fought briefly. As I maneuvered to get behind him, I saw Flt Sgt Scott achieve a few strikes on the engine. I completed my roll, pulled a high G turn and quickly got behind him as he tried to turn on me. With another burst I shot him down in flames.

Returning to the fray I engaged another short nosed Fw190 as he was in a climbing turn. More through instinct and an ardent desire to prevent these marauders from inflicting harm on my compatriots than deliberate decisions and choices, I maneuvered in behind him where I was able to shoot him down with another burst. Squadron Leader Shephard claimed sharing in this kill as he was apparently below the enemy aircraft shooting at the same time. I never saw the Squadron Leader. The Fw 190 burst into a mass of fire just behind the engine and he went down in flames. We were just a little closer to ending this madness of war.

In all we shot five of them down with no damage to any of our aircraft. The others fled. I discovered two interesting things in this aerial engagement. One is that more than fifteen aircraft were tumbling about in a small space in the sky and yet there was never any danger of mid-air impact due to the acute situational awareness of each of the pilots. The other interesting observation is that the adrenaline levels are so high that time seems to be stretched out giving one what seems to be longer than normal time to possibly plan and employ mostly instinctive actions to execute successful combat maneuvers.

Outnumbered but Victorious

Our daily battle planning consisted of our pilots gathering around a map table with the latest after-action and intelligence reports. With the aid of the intelligence officer, and taking into account the enemy's activity, we determined where we could do the most damage to the enemy's supply lines.

Late in the afternoon 28 April 1945, five of us stood around the strategic maps in the briefing room at Celle. Four were pilots and the fifth was our intelligence officer, Flt Lt Thomas Weston Peel Long Chaloner, The Honorable Lord Gisborough, 2nd Baron Gisborough of Cleveland, Yorkshire, a WWI pilot and ex-Prisoner of War who returned to RAF service during World War II. He served as 41 Squadron's Intelligence Officer for over five years of the War, and reported the squadron's activity, victories and losses up the chain of command on a daily basis. He refused further promotion. *Source:*

https://en.wikipedia.org/wiki/No._41_Squadron_RAF.
We affectionately referred to him as "Gizzie."

Gizzie was a most likable and as you can imagine, a well-connected fellow, with curly, graying ginger hair and curly side-whiskers down to his chin. He had some pretty tall stories to tell from World War I. One was about fighting a Hun in their open cockpit biplanes. When they both ran out of machine gun ammunition they shot at each other, side-by-side, with hand guns until they ran out of ammo and then ended up throwing their handguns at each other before heading for home. On a number of occasions, the authorities attempted to have Gizzie posted away from the squadron. Whereupon Gizzie would go up to London, meet with some Air Marshals and the posting would promptly be cancelled. He was with the Squadron throughout the whole of World War II. We were in Denmark after the cessation of hostilities when Gizzie was demobilized and we flew an honor guard accompanying his DC3 half way across the North Sea.

As the five of us studied the maps, intelligence reported that the Luftwaffe would take to the air at dusk to try to avoid our fighters. Consequently, late that afternoon a flight of four took a wide sweep from Celle, where we were currently stationed, to the Baltic Sea. We carried 45-gallon belly tanks to give us extended range.

As the sun began to sink slowly in the west, we flew around 30,000 feet well out over the Baltic so the Germans would think we were continuing in that direction. Then we circled back over the Baltic coast and cruised inland at about 25,000 feet. Approaching the town of Schwerin, we spotted about ten or more aircraft at low level and headed down toward them. Then something curious happened. We saw explosions around the airfield and town, so two of our number assumed that they were RAF Typhoons attacking the area and climbed back up to our cruising altitude and headed home, but Flt Lt Tony Gaze and I continued on down to investigate further.

Spitfire XIV with 45-gallon belly tank

As we got low enough to positively identify the aircraft we realized they were Fw190's, a large number of them. We assumed, incredibly, that they had spotted us and were jettisoning their loads in preparation for a fight. Diving down from our cruising altitude, we had built up some excess speed. I picked out the highest Fw 190 flying quite slowly. I had to lose speed rapidly using my propeller in fine pitch as a brake and fish tailing as hard as I could to avoid over shooting him and becoming the hunted versus the hunter. I was very close to him when I opened fire, without knowing what the round object under his belly was. I soon found out as I fired with cannons and 50 caliber machine guns.

The round thing was a bomb and there was an almighty explosion. I ducked down for maximum protection from my bulletproof windshield and large engine. Although the outside air was very cold, I could feel the fiery heat on my neck between my collar and cloth helmet. I could see the inferno of blazing fuel and rubble washing over my wings and engulfing my Spitfire. As I weaved through the flaming wreckage I decided it was time to take stock of the condition of my aircraft. I was still flying so I attempted to gain a more defensive altitude, but I was obviously very badly damaged. I was vibrating so badly due to a damaged propeller, I couldn't open the throttle too much or I would pull the engine from its mountings. In addition, I was wallowing so I knew my tail was badly damaged. Now my concern was whether the debris had punctured my radiators, one under each wing. However, my engine continued to run without overheating. My other concern was being unable to adequately fight and the possibility of being picked off by any of the Fw190's I left behind. I climbed using as much power as I dared to gain all the altitude I could get as a safety factor. I had perhaps a hundred miles back to our base. So I called for a homing to take the most direct route back to base. The radio direction-finding personnel were on the ball and got it to me immediately just before the Germans, who were listening in, jammed the radio.

As I was dealing with my predicament Tony got entangled with the group of Fw190's, popped into a cloud, spun out and headed home. He did tell me however, that an Fw190 was either knocked down or destroyed on the ground by the debris from the Fw190 I shot down. My escort home was the ultimate shepherd, my Lord and Savior, who kept me airborne and free from attack.

Before getting too low on approach to the airfield I tested my flaps and undercarriage. Both were still functioning. So with the crash crew standing by I came in fast in order to retain control until my wheels were safely on the ground. After climbing out of the cockpit, I discovered that a piece of one blade of my propeller had been split off long ways, paint was burned off the wings and fuselage and part of the controlling surfaces of the tail were missing, plus I had acquired numerous holes and dents in the wings and fuselage. Most remarkable of all was that nothing entered the huge radiator and oil cooler air scoops under each wing, even though the narrow cowling edges of the scoops were riddled with holes. The photo of the propeller adorns the cover of this book. Make no mistake: The hand of the Lord was indeed upon me. Had the natural course of events been allowed to proceed I would either have bailed out or made a belly landing in Germany with the alternatives of becoming a POW or being killed. Neither occurred – I came home on the wings of the Lord. This was no coincidence, the normal course of events had been altered as it can still today to those who turn to Him. Think about it.

My damaged Spitfire resulting from an exploding Fw190

The total number of aerial victories in which I had participated was now at five Fw190's and one V1 Flying Bomb, some solo, some shared. It was customary to share portions of victories between pilots when multiple aircraft were engaged with the same target. I remain unconvinced however that this practice was always fairly implemented and I further believe it may have been used to advantage by some who had ulterior motives. Certainly the squadron had no intention of inflating aerial kills but a proper method of victory distribution had to be devised so all pilots would receive appropriate recognition while at the same time correctly accounting for downed enemy aircraft. I heartily believe that I should have been credited with more full victories than the record indicates. The official historical record shows that I shared one half in the V1 kill on 11 August 1944 and one half each of three Fw190's in April of 1945, along with two full victories, or total of four victories when the sharing arithmetic is finished. There is room for interpretation when one gets down to looking at the aerial engagements closely. It is my firm conviction that I was singularly responsible for the destruction of the V1 on 11 August 1944 and the Fw190 on 28 April 1945 for which I received half credit. I believe the second Fw190 was destroyed as a result of the bomb exploding on the first. Using this approach my total victories should have been five. I will remain convinced of that. At 22 years old and somewhat reserved in my ways while on the ground, I was not given to argue. While in the air however, with the help of my Lord, I felt I had few peers.

John F. Wilkinson

Bücker 181

When the Germans pulled out of Celle, they did so rather hastily. They left quite a few aircraft scattered around the airfield and in the woods. We could hardly understand how they got some of the aircraft between the trees they were jammed in so tightly. There were fighters and all manner of single engine machines. Rather than leave us any that we could use, they smashed them with sledgehammers and set fire to some of them.

While we were actively striking daily at the enemy and fighting with the Luftwaffe, we did have some time to spare. So my friend and fellow fighter pilot Flt Lt Fanny Farfan and I set about keeping ourselves interestingly occupied during our spare time. Fanny was from Trinidad (don't laugh at the name, he was all man!). My friend Fanny and I looked at the wreckage and an idea was born. Wouldn't it be fun to build our own airplane? So we set about finding enough pieces from the wreckage that looked as if they would match up closely enough to make one aircraft. We collected a wing here, a tail there, part of a fuselage somewhere else. We found wheels and struts and an engine that had been spared the sledgehammer. We removed the propeller from one wreck and raided the German storage hanger for instruments, pitot tubes and other bits and pieces. We even found a whole canopy. Rudder bars, control wires and tools were also collected. All of this we dumped in a disused hanger.

Since there was a certain amount of manual strength required, Flt Sgt 'Jock' Stevenson agreed it would be fun to help. It was with the assistance of Steve Brew's book "Blood, Sweat and Valour" that I was able to correctly identify 'Jock" Stevenson. Gradually, over the next few weeks our machine began to take shape. We installed dual controls and two full sets of instruments, giving it all weather flying capability. Owing to our good fortune that we were well trained in engines, airframes and aerodynamics we were able to fit the various pieces together properly. We had fun doing this. Since it was basically a German flying machine we thought it best to paint it yellow, the color of training machines, and paint the Royal Air Force roundels on it to identify it as one of ours. A necessary precaution, wouldn't you say? We 'borrowed' a spray gun and a few gallons of yellow paint and went to work. The job of spraying the underside of the plane fell to 'Jock', unfortunately for him. Something went wrong with the spray gun and he was drenched in yellow paint. He was rather odd looking for several days, even more so than some of today's teens!

While all this work was going on, the Engineering Officer took a very dim view of our efforts and our creation. After checking everything and running the engine, the day finally came when it was ready to take the air. To avoid attention, we flew it at dawn and it performed beautifully. The

Our yellow Bücker 181

Engineering Officer was furious and went to the Group Captain to complain. He had disapproved of the project from the time he first heard about it. The Group Captain called the three of us into his office and we figured we were in for it, whatever 'it' might be. With a stern face he said, "I understand that this morning you flew the plane you have been building. The Engineering Officer has made a complaint on the basis that he has not inspected it and disapproves of the liberties you have taken." Then with an easing of his stern attitude he said, "I had to call you into my office for appearance sake, but I think it is a jolly good show." In addition, with that he dismissed the whole complaint.

Later we found a German aircraft recognition book. Our machine was a Bücker 181, a two seat light plane, and the book showed that ours had a better performance than factory built machines! We flew it with us every time the squadron moved.

In war, we are constantly reminded that the Lord is always watching over us. As a result many miracles occur. I heard a report that during one of the bombing raids over Nazi occupied Western Europe, a bomber was destroyed by anti-aircraft guns and on fire. The pilot instructed the crew to bail out from the blazing inferno inside the bomber, which they did not hesitate to do. Alone in the aircraft the pilot found that his parachute was on fire and useless. He was faced with the choice of two courses of action. Stay and be burned alive, or jump to his death without a parachute, neither a pleasant prospect. He preferred the less painful alternative and decided to jump, sans parachute.

The human body, falling through the air, reaches a terminal velocity of about 125 miles per hour. As the pilot hurtled earthward, he fell through the branches of a tree and landed in a soft snowdrift that broke his fall. Incredibly, he was able to get up and walk away to spend the rest of the war in a prison camp.

Low Altitude Parachute Deployment

A member of my squadron was the recipient of a similar miracle. Having flown earlier sorties that day, I was not on this particular sortie but I heard all about it. It was dusk, as Flt Lt Terry Spencer led a flight up to the Baltic Sea where they spotted some transport ships in the lee of the coastline. They wheeled around and came in at 400 mph for a low level cannon attack. However, in the darkness, they had not noticed a destroyer in the shadow of the cliffs. Terry was now just 30 feet above the water when the destroyer opened fire and scored a direct hit on his Spitfire with a heavy caliber projectile. The explosion broke his straps and blew Terry out of the closed cockpit. He later said he saw the water racing up at him and he figured it was the end, but then his parachute streamed out behind him and broke his fall into the water. He struck out for the shore swimming mightily when he stubbed his toe on the bottom! He had landed in a shallow area beside the shipping channel.

He was taken prisoner of course, but he reported that the Germans gave him a certificate verifying the lowest successful parachute jump on record. He was taken to a prison camp near the front lines. As our army advanced, the prison guards deserted the camp. Terry and another pilot stole a motorbike that they rode back to our lines. He was soon with us flying again, thanks to the Lord.

Our squadron motto on the squadron's crest was "Seek and Destroy," which we fulfilled to our utmost and continued to do so while we were deployed. Sometimes

John F. Wilkinson

we replaced the motto with, not so humorously, *"dicing with death."*

While at Celle we devised a strategy of flying at high altitude at dusk to the Baltic Sea, turning around and letting down to lower altitudes to catch the Luftwaffe coming up for night time missions. It led to some interesting engagements and beautiful sights.

There is wondrous beauty and many grand sights to be seen and appreciated in the skies of our world, not readily seen as passengers in an airliner. One of these I observed in the cold clear predawn skies as we took off from Celle into the magenta colored sky. As the pilot taking off ahead of me opened the throttle of his Spitfire, long blue flames poured out of the twelve exhaust stubs down the sides of his machine and reached perhaps as much as 50 feet beyond the tail surfaces. This phenomenon could only be seen at night. It was a remarkable sight as each of us lifted into the early dawn sky. This illumination is the primary reason Spitfires were not often flown at night. Those flames made them easy targets and the pilots could not see adequately over the flaming exhaust flowing past each side of the cockpit.

Part V: Post War Lessons & Antics

THE DAY AFTER WORLD WAR II ended in the European theater, 5 May 1945, 41 Squadron was further deployed to Copenhagen, Denmark ostensibly as a warning to the Russians - hands off. We were there for defensive purposes where we spent three wonderful months. We were still flying Spitfire XIV's. On our way up to Copenhagen we couldn't miss seeing curious circles in the Danish farmer's pastures. No, they were not the strange crop circles we now read about. These circles proved to be the farmer's ingenious method of feeding more cows in any given field. The cows were tethered to a row of stakes driven into the ground in a line across the field. When all the grass was consumed to the length of the tethers, the stakes were moved up thereby allowing the grass to grow behind them and feed more cows per acre.

While in Copenhagen we were stationed at Kastrup aerodrome. Kastrup was the premier Copenhagen airport with its runways extending to the edge of the Baltic Sea. Very comfortable accommodations were provided for us there. We received a very warm welcome from the Danes who were now released from Nazi tyranny. At that time Denmark was considered the larder of Europe. After many months of subsisting on dried and canned foods the fresh foods now available were a joy to us! I drank a whole pint of their richest cream straight away.

Some weeks after our arrival at Kastrup we were advised that General Montgomery would soon arrive to meet with the Danish Prime Minister. The Prime Minister met General Montgomery at the Kastrup airport. For this occasion the British forces planned a Victory Parade to proceed through

Flying the Auster over Copenhagen for the Montgomery victory parade

Copenhagen. I was assigned the task to cover the parade from the air in the Taylorcraft Auster, a small high winged monoplane used for liaison. I was to take a reporter and a photographer and ordered to take them wherever they wanted to go, no restrictions. An order I carried out to the letter! When the parade got underway I took off with my passengers, the photographer with his camera sat beside me and the reporter in the back seat. Both were from Copenhagen's premier newspaper, the Politiken.

The weather was warm with broken clouds and good visibility. Starting from a few hundred feet up in the Taylorcraft, I glided down over the parade to a few feet above the rooftops, banking for the photographer to get good shots before turning away and climbing up again trying to be as considerate as I could with regard to engine noise. Since telescopic lenses were not available at the time this enabled the photographer to get some excellent photographs of the

Victory parade in Copenhagen for General Montgomery

parade. This process was repeated a number of times until the parade reached the main square where it terminated.

The next day it was brought to my attention that General Montgomery was considerably displeased by my presence above the parade. But since I had done no more than obey orders his displeasure did not come down on me. There was a big write up in the newspaper and the photographer gave me some very nice prints. One print taken by the photographer was of the reporter being sick in his camera case!

Sometime later I was asked to perform a solo aerobatic display for the dignitaries and visitors. One of my "stunts," not performed by any other pilot that I know of, was to take off and complete a slow roll before passing the other end of the runway; a feat I would not attempt in any aircraft other than a Spitfire XIV. I did keep enough of a margin of airspeed so that should the engine fail I could straighten out and make a belly landing in the shallow waters of the Baltic just off the end of the runway. However, the Rolls Royce Griffon 65 was incredibly reliable and never failed me.

Climbing to a reasonable altitude over the beach I proceeded to perform every aerobatic maneuver I could think of. One maneuver I had never seen performed by a four ton fighter was a 'falling leaf'. This is quite tricky taking careful timing and lots of working the controls in order not to 'fall out of the sky'. I would reduce speed to a stall, start a stalled spin to the left and immediately shift to a stalled spin to the right and back and forth that way until time to climb up again. This gives the effect of the fighter floating down like a falling leaf. It really was fun.

After an hour and a half, and a full tank of fuel, I returned to the airport. Circling around the pattern, I made a normal curving fighter approach to landing, except upside down with wheels facing up. At the last moment I rolled over, lowered the flaps and made a perfect three point landing. I was bathed in sweat and so exhausted I had to leave my parachute in the seat for the ground crew to bring it in for me!

Aerobatics and Close Formation Flying

After four years of war, during which I had experienced the indiscriminate bombing of London and a year of combat, Copenhagen was like a three-month stay in heaven! While there my friend and fellow pilot Fanny Farfan and I decided to pioneer close formation aerobatics flying in our Spitfire XIV's. The Spitfire XIV was a four ton single seat fighter with a 2,050 horsepower Rolls Royce Griffon 65 engine and a big five bladed propeller. After some experimenting Fanny and I had the procedures worked out and trained two of the most proficient pilots to join us. It was lots of fun for us using our skills honed in battle. As we practiced, we had one main rule; if something didn't feel just right we were to break apart.

There were such occasions in the early stages and we were told by observers on the ground it looked as though four fighters were suddenly falling out of the sky. We were being very careful for obvious reasons. Then we would form up and try again until we perfected it and worked out a complex schedule of maneuvers. Even today, when watching close formation aerobatic teams, I see they use the same maneuvers we had pioneered, but of course much faster with jet powered aircraft.

The leader of the aerobatic team is required to fly extremely smooth and precise maneuvers with minimal power changes. When power changes are required they are accomplished with smooth throttle adjustments to permit continued precise positioning of the other members of the team in relation to the leader. The other members of the team must concentrate solely on the leader with no concern as to up, down, where the earth is or what maneuvers are being performed. That requires considerable skill and microscopic adjustments of throttle, propeller pitch, constant stick and rudder movements in order to constantly maintain a perfect position relative to the leader. However, for the skilled pilot these factors are somewhat secondary stemming from the experience of long hours of combat aerobatics. The application of control has become a reflective response to what the pilot wants. In this case you could say the wing and slot aircraft responses follow a path from the pilot's eyes to his brain to the physical application of controls in a seamless transition occurring in microseconds – a clear manifestation of man and machine operating as one. Likewise, the image of all the instruments with needles in their proper positions becomes so ingrained in the brain that only the most fleeting glance at the overall instrument panel immediately alerts the pilot that something is wrong if the indicators are not correctly positioned.

For our finale we pulled vertically straight up, splitting into four equally spaced directions, then with timed precision curving down to almost ground level, leveling off and heading straight for the center of the airfield. We intersected stacked four deep as we continued to the four points of the compass. From there we wheeled around to land in line astern, or if the runway was wide enough, in pairs in formation. In wartime we would take off from our grass field as many as four together. This was necessary to get a whole squadron in the air quickly to conserve fuel for long escort missions.

I concluded from my personal experience and analysis of close formation flying in Spitfire XIV's (or most other single engine propeller driven aircraft) that close formation flying requires considerable flying skill, training and practice and even then the risks are far greater than flying solo. In addition formation flying is terribly unforgiving of the slightest lapse of attention and it is made even more difficult by the massive amount of torque developed by the Griffon 65 engine and five bladed propellers. On a formation take off, one must compensate for the increased torque. This torque is partially alleviated by the design of the vertical stabilizer of the Spitfire XIV that is slightly more offset than other Spitfire Marks. Close formation flying with Spitfire Mark XIV's calls for some of the qualities gained by experience and in our case from one-on-one fighting to the death in close aerial combat. A full measure of self-confidence and the drive to be the best are imperatives, as well as the need for a fine sense of situational awareness. Fear does not enter the equation of the well trained and experienced fighter pilot as the drive to be the victor, the constant flow of adrenaline and the need for concentration overrides any hint of fright. Fixation on single tasks that require separate thinking and analysis are the harbingers of disaster. Another essential ingredient is possession of the determination to push machine and body to hitherto unknown limits when necessary. These latter points are altered somewhat for close formation flying where there is a requirement for complete concentration on the leader and to do only that required to maintain a fixed position relative to the leader.

John F. Wilkinson

After three months of living in the lap of luxury in Copenhagen we were re-deployed to Husum in Schleswig-Holstein, Germany for a few weeks before moving to Lübeck, Germany, on the Baltic Sea. Lübeck was right at the border between Allied territory and Communist Soviet territory, separated by a canal. On our side of the canal one might see the occasional farmer plowing his fields. On the Soviet side of the canal there was a soldier placed every few hundred yards each armed to the teeth. The guards, it seemed, were to keep the Soviet dominated peoples from escaping, rather than to keep us from entering. Who in his right mind would have wanted to!

When I was young my mother and father were close friends with two other couples. After my father died both the husbands of the other two couples died peculiarly. Of the three widows they expected my mother would be the first to remarry because she was the prettiest. In fact she was the one who never remarried. One of her two widowed friends married a Dane and moved to Denmark with him before the war. After my squadron was moved to Copenhagen my mother sent me the name and address of her friend in Denmark. It so happened they lived on the north end of the same island as Copenhagen. Somehow I managed to get in touch with them, checked out a motorbike from the RAF motor pool and drove up to visit them. They were a very warm family with an attractive upper teen-age daughter. I might add that during the war they lived in constant fear the Nazis would find out she was English born and send them to one of the death camps. Occasionally, while stationed in Copenhagen I enjoyed visiting the family. After moving to Lübeck, occasionally when I was assigned to a particular duty in

Copenhagen, I would prepare a note folded around some cigarettes. (I did not smoke but some of my friends did.) I would fly slowly with flaps down, at low level over their house and toss the package into their back garden. This let them know I would soon be on a motorcycle coming up to visit them. On one occasion I decided perhaps I might be a little too low. Not wanting to take any tiles off the peak of their roof I rammed the throttle wide-open as I have done many times. However this time I blew a cylinder head. Choking fumes filled the cockpit. Not planning to fly at high altitude I was not carrying any oxygen. Very carefully and very cautiously with an open canopy I took the hose from the end of my face mask and edged it just a tiny bit to the edge of the cockpit to allow fresh air to enter my mask. To move it any further would have blown the mask off my face. In that way I flew back to Copenhagen where I had my engine repaired.

Air Show in Copenhagen

After we moved to Lübeck, news came to us that there was to be an air show at Copenhagen and 41 Squadron was to be featured with our aerobatic team as the main aerial event. This was to be a grand air display for the Danes at Kastrup Airport. However, on the day we were to leave a front came in and the weather turned sour with heavy rain and dense low clouds preventing us from flying north the day before the air show. The squadron was grounded with the hope that we could fly to Copenhagen the next day and take part in the air show without landing. After some discussion it was agreed our aerobatic team would fly up through the clouds and fly to Copenhagen so we could keep the crowds entertained until the rest of the squadron arrived later in the morning. The weather rarely stopped us during the war!

Our aerobatic team consisted of the finest and most experienced pilots. Since we were billed as the main event, we just couldn't disappoint the crowd! Accordingly, at dusk the four of us took off and climbed up through the rain and clouds. At dusk we cruised in battle formation on top between the unusual towering columns of clouds. With a dark blue sky above, the billowing clouds, tinted by the setting sun with a broad spectrum of pastel colors formed an ever-changing kaleidoscope of beautifully multicolored columns. It was so calm and peaceful with the smooth hum of the engine adding a muted symphony to compliment the scene. It was an experience forever etched in my memory.

As we flew north between the columns of clouds the drone of the engine faded into the background. It was as if one were floating on a magic carpet. The beauty of the clouds and the skies as we flew through the calm air above the storms offered one of those times that touches the soul and is long remembered. My senses were alive and keenly attuned to nature's beauty. It gives one a greater appreciation of what God hath wrought.

The next morning as we went to flight-check our machines we noted a huge crowd of Danes was already gathering. We later learned there were about 300,000 spectators eager to watch the show. A newspaper reporter from the Politiken Newspaper, now a friend of mine, was allowed to climb on my wing and examine my cockpit. We taxied out, in line astern, to the end of the short runway. As I started my take off run I rapidly increased the throttle to full power (unusual due to heavy torque). The moment my wheels left the ground, I raised the gear, staying low to quickly gain the speed I needed to perform my favorite slow roll before crossing the other end of the runway that terminated at the beach and the Baltic Sea. After that we individually performed all manner of aerobatics, saving our close formation aerobatics for the assigned time slot.

In preparation for the air show a collection of old trucks and other objects had been assembled on the beach at the end of the runway. When the rest of the squadron arrived, they came in waves with guns blazing, strafing the vehicles on the beach with 20 mm cannon and 50 caliber machine guns. The cannons fired explosive shells while the 50 calibers launched armor piercing and incendiary projectiles that exploded and burned the assembled targets. All in all we put on quite a show.

Eggs in the Belly

Following a few weeks at Husum, Schleswig-Holstein, Germany, the squadron was sent to Lübeck on the German Baltic coast. So, it was back to dried and preserved foods and the two or three week rebellious tummy schedule.

In time of war, food is always a problem. As Napoleon quipped, "An army marches on its stomach." Food was a problem in England and even more so when my squadron moved across the channel to continental Europe. Most of our food was dried or otherwise preserved and had to be reconstituted. Potatoes were of the consistency of Elmer's glue and eggs had all the characteristics of dried granular yellow paint. The stomach could take it for only two or three weeks before rebelling, fortunately only temporarily. One had to spend time on a seat quite dissimilar to those in our cockpits.

It is said that necessity is the mother of invention, and so it was in this case. I began to think about our recent experience in Denmark. Since Denmark was the larder of Europe at the time, perhaps there existed an opportunity. After some thought, I proposed that anyone who wanted fresh eggs for breakfast could put up the money for them and I would get the eggs. The money and the orders flowed in from ground crew and pilots alike.

Now the cockpit in a Spitfire is very tight quarters with barely room for the pilot. Big men could not fly Spitfires. So there was no way to transport food in a Spitfire. However, as you have probably guessed, I had an idea. When flying over enemy territory during the war, we used to carry 45 or 90-gallon belly tanks. There were still plenty of the 45-gallon belly tanks around. Some of the tanks were metal and some were made of treated wood.

I asked the ground crew to bring out two of the unused wooden 45-gallon belly tanks. We broke away the wood on the top of the tanks that fit against the belly of the aircraft. Then we assembled them under the bellies of two Spitfires. I chose a pilot from the squadron who was known for his smooth landings and together we flew up to Copenhagen. Since the Royal Air Force was still in residence at Kastrup airfield, I was able to commandeer a fifteen Cwt. truck (Hundredweight – equaling about 112 lbs. In the US we'd call it a ¾ ton truck) and drove into town. Eggs were plentiful in Copenhagen and so it was there that I purchased 1,500 fresh eggs. That was quite a sale for the shopkeeper.

Back at the airfield my fellow pilot and I very carefully filled the two belly tanks with eggs. We did not break one of them. With the assistance of some ground crew the tanks were reloaded back under our Spitfires. Then with great care we taxied out to the runway and took off, hoping and praying that our tanks did not fall off. Landing with even more care when returning to Lübeck, we arrived safely with both tanks still in place.

The mission was not without its risks. I remembered that when we were in Holland I was on a solo mission and returned with a full 45-gallon belly tank. As I banked around onto final approach to land I felt a slight bump. Just as I approached the end of the runway the tank ruptured and sprayed 45 gallons of 150 octane fuel on the grass between the runway and the fence. That, in and of itself, was of no major concern. It became a delightful event, so I was told by controllers, for a little Dutch boy who jumped the fence and ran over to collect a piece of the broken tank - a prize for his memory box. I didn't want that to happen with these egg loaded belly tanks!

Eggs in the belly tank – catching up on our dining after the war

The ground crew gingerly removed the tanks and eagerly removed the eggs. Some of them got a bit too eager, but only half a dozen were broken. So Humpty Dumpty's fall was a lot better than it might have been. It doesn't take much of an imagination to visualize what might have happened had the tanks ruptured or separated from the aircraft, especially when landing!

Oh, what joy! We had fresh eggs for breakfast and a momentary break in the now rather boring daily routine and even more boring and distasteful menu.

Yachting on the Baltic Sea

While in Copenhagen, our Group Captain laid claim to two sailing yachts formerly owned by the Nazi invaders. One large yacht had two masts and the other, not quite so large, had a single mast. He also commandeered a German torpedo recovery launch. The launch was big and powerful with a crew of two German seamen to maintain and run it. We were happy to have them run it and they were happy to comply since they were provided with food and lodging. The launch could sleep several and was equipped with a good kitchen.

The smaller of the two yachts was still quite large by most standards, since it had berths to sleep six. My friend, Fanny Farfan, was raised on sailing boats. He was an expert yachtsman with whom we later enjoyed sailing the Baltic. He was also as strong as the proverbial ox. On one occasion while sailing up the channel to a dock, we became firmly stuck on one of the shifting sand banks. No problem for Fanny. He just jumped over the side, got under the yacht and pushed it free of the sand bank.

After we were well established at Lübeck, the Group Captain decided he would like to have the smaller of the two yachts berthed close to Lübeck and available for his use. Knowing of Fanny's yachting capabilities, he commissioned Fanny to sail it down from Copenhagen. Fanny immediately set about laying his plans.

Fanny gathered four volunteers from among our pilots to form the crew for the cruise down through the Danish islands and across the Baltic. Not one of the four had any sailing experience but that did not deter them.

I agreed to keep a daily check on their progress from the air. Fanny and his crew gathered all the necessary equipment and supplies for the trip. He then made arrangements for a twin engine Oxford transport to deliver the supplies to Copenhagen's Kastrup airfield.

The best laid plans, as the saying goes, can go awry, and "Murphy" was busy derailing them. For some reason the Oxford never arrived but the dauntless crew was not to be deterred. They pooled their money to buy food and other basic needs. They had some charts, but they did not have a compass, so they removed one from a partially destroyed German aircraft and departed on schedule.

I had two Spitfires up at Kastrup that would be returning to Lübeck, one on the day following the departure of the yacht and the other on the next day. So I detailed the pilots to spot and check on the yacht's progress. All was apparently well and as it should be (although we did not know that the Oxford had not delivered the supplies). There was no telephone contact between Copenhagen and Lübeck at that time and a portable radio transmitter was not available.

On the third day the weather had deteriorated very significantly as I went looking for them myself. It was very stormy and the clouds so low I could not fly very high to gain a wider field of view. I searched their intended course and not finding them I widened my search until shortage of fuel forced me to return to Lübeck.

On the fourth day I took our Spitfire XIX, a photo reconnaissance version with extra fuel capacity. The weather had improved but there was no sign of them until I spotted the yacht in the harbor of a small Danish island. I did not see any sign of the crew. When I returned to Lübeck it was too late in the day to make the trip in a light aircraft, so it was not until the next morning that I taxied out in a Taylorcraft Auster, a small high wing monoplane. I flew up to the little island and found a farmer's field that I proceeded to drag. Dragging a field means flying low and slowly to inspect it for holes and rocks, etc. with a view to landing. It looked very good and so I landed safely. Immediately a crowd of Danes appeared as if from nowhere. I climbed down and was glad to find that some of them spoke English, because I certainly could not speak any Danish.

The people surrounding my plane did not know anything about the yacht in their harbor. So I detailed one responsible looking gentleman to watch the aircraft while I went to make some inquiries. The Danes are a very hospitable and friendly people and their island telephones were working, so one of them helped me to phone the Danish Coastguard who also spoke English. I asked the Coastguard about the yacht in the harbor and told them the reason. It turned out that I had found the only sister ship to the one I was seeking. The only thing the Coastguard could tell me was that a yacht had been seen in the storm moving or drifting toward the Russian zone. This was not good news! I returned to the plane and gained the assistance of a man to stand on the brakes while I swung the prop. There was no self-starter in the Auster and I was an experienced propeller swinger! So off I went back to Lübeck with my bad news. That night I took the pilot parachute out of a discarded German parachute. The pilot 'chute is small and designed to deploy the main parachute. I prepared a message in case I found the yacht and put it in a closed can that I attached to the parachute. The message instructed the yacht's crew to find a good landing place for me and that I would fly up in the Auster to find out what had happened and what they needed.

The next day I took another Spitfire with me to assist in the search. We covered a large area of the Baltic and the Danish islands between Copenhagen and Lübeck. I finally spotted the yacht in a small harbor on the south-eastern edge of the main island that Copenhagen was on. When I circled the crew came running down to the jetty waving like mad. I flew a little way out to sea, slowed down, lowered my flaps and opened my canopy. I made a very low slow approach to the harbor and calculating the right moment I threw out my tin can with its parachute. The 'chute opened and the can landed right on the jetty. Bull's eye! With no means of communication they had to wait until I found them. So, with a final wave it was back to Lübeck to lay more plans.

The following morning saw me taking off bright and early in the Auster with extra fuel cans in the back. I flew up to the island harbor where the crew of the yacht await-ed my return. As soon as I circled they came running out and waived for me to follow them up the road. Of course I could not fly slowly enough to stay with them and had to circle in order to follow them, when suddenly they waived me away because, I learned later, they thought I was going to land on the road. This was rather confusing, so I spotted a field that might be a possible landing site. I circled down and lowered my flaps for very slow flight to take a close look at the surface. I decided it was not suitable and opened the throttle to climb away when with a bang the flap lever snapped out of its slot and let the flaps up. This meant I could not gain enough speed to climb. A most undesirable situation since in front of me I was faced with a telephone pole, wires, a large tree and a fence between them. I had no choice but to fly under the wires, miss the telephone pole with my left wing by inches and thread my right wing between the branches of the tree. Whew! I made it and climbed up out of the next field with barely an inch to spare. Once again the Lord saw me through what by normal scenarios would have been a complete disaster resulting in my demise. "Well, enough of that," I said to myself, "I'll go farther afield." So I found a nice looking field a short distance from the village and landed safely. The crew came hurrying up to where I had landed, along with the obligatory spectators, and I learned their story.

The pilots on board the yacht set out on this new adventure in high spirits. On the first day they found that the liquid in the compass from the German aircraft had leaked out. But not to be defeated, they took it apart and filled it with kerosene which worked just fine. By the next day they were getting into the swing of things and the inexperienced ones were learning what was required of them to handle a yacht of this size. But, as we know, the weather on the third day became disastrously wild. One after another their sails were ripped and became useless. Even the storm sails suffered the same fate and they were being blown toward the Russian zone where they would have received a most unpleasant reception! There was no auxiliary engine. The boat and all their supplies were swamped and they were bailing to keep it afloat.

So a bit of ingenuity was called for. Among the supplies they had a sack of sugar and a tin of sardines. They took the key for opening the tin of sardines and broke the handle off, providing them with a crude needle. Then they unraveled the sugar sack to provide them with thread and stitched up one sail. Fortunately, the worst of the storm blew itself out quite quickly. With their patched up sail they managed to make it back to a Danish harbor where they had to wait until I found them.

After my scary arrival, the yacht's crew and I sat down and made out a lengthy shopping list that included a Very pistol and signal flares, and money to pay for the work on the sails that a local sail shop was repairing. After all this was completed I refueled the Auster from the cans of petrol and returned to Lübeck. On reporting to the Group Captain, he provided the money and ordered all the necessary supplies that were duly loaded into the Auster. Along with the extra cans of petrol, it was quite a heavy load for that small aircraft and so it meant that I would have to land on one of the islands to refuel before reaching my destination. This time I found a small deserted island with a wide-open grassy plateau. I dragged it and landed safely.

But now comes the tricky part. The Auster can only be started by manually swinging the propeller, not a wise thing to do without someone in the cockpit. There was one incident when a pilot attempted this and the plane took off by itself without him! Now the fuel tank in the Auster is between the engine and the windshield. It was necessary to stand on the landing wheel, lift the petrol can and pour the petrol into the tank with the engine running therefore blowing quite a bit of fuel against the windshield, and some on me too. However, I did get enough petrol into the tank and was soon on my way again. When I landed at the harbor, the crew had borrowed a horse and cart to carry my load to the yacht.

I had to admit that there was some nefarious reason for being in that particular harbor. It turned out that the local baron owned most of the nearby land and that his daughter was Fanny's girlfriend. She was visiting him at the harbor! The baron was very wary of military personnel. It was now late afternoon.

Fanny asked me to fly his girlfriend to her father's baronial home with the promise that I would take the baron for a flight. I agreed to this and was invited to stay the night with the baron. So I got the baron's daughter strapped in and we prepared to leave. She was rather buxom

Ready to fly the Auster

and I did not grasp what a big, heavy girl she was. I taxied to the extreme end of the field, the same one I had flown out of the day before, set the flaps, stood on the brakes and ran the engine up to full power before releasing the brakes. We trundled across the field, the extra weight keeping us from gaining speed as quickly as we should have. I had to make an instant decision. Chop the throttle and run into the hedge or go for broke. I chose the latter. This all occurred in a microsecond as we careened across the field. In front of me was a hedge. It had a gap in it that was closed by a wooden fence. In the middle of the fence was a tree stump about five or six feet high, apparently the reason for the gap in the hedge. On my side of the fence there was some long grass. Upon reaching the grass and still without flying speed, I pulled back on the stick forcing the plane into the air in a stalled condition, then ramming it forward forcing it to hit the ground and bounce high enough to clear the fence with the tree stump between my wheels. Then down into the next field for a 'normal' take off. Such maneuvers cannot be planned but are somehow executed instinctively. Once again it was the hand of the Lord guiding me to react in a way that would save me and my passenger from a deadly situation. I gingerly looked out of the corner of my eye at my passenger to see if she was white with fright, but she was sitting there 'fat, dumb and happy' as though everything was normal. Of course, I did not enlighten her. We landed in a nice smooth field by her father's mansion.

After dinner, a pleasant visit with the family and a good night's sleep, the baron and I got up early and I flew him around to see his lands. He thoroughly enjoyed it and Fanny told me later that from then on he was very friendly to our pilots. Upon returning to Lübeck early that morning I arrived just as a section of Spitfires was preparing to take off to come looking for me. The rest of the yacht's cruise was smooth sailing and I circled them each day until they arrived at the Travemünde harbor near Lübeck.

John F. Wilkinson

Farewell to the Spitfire XIV

While at Lübeck on 15 September 1945 we bid farewell to our beloved Spitfire XIV's and we were reequipped with the Hawker Tempest V's. The Tempests were a little bigger than the Spitfires. They boasted a more powerful 2,180 horsepower, 24 cylinder H arrangement, Napier Saber engine and a big four bladed propeller. With that many cylinders compared to the twelve cylinder Spitfire Griffon 65 engine it was so smooth to the pilot it sounded like a sewing machine. It was a delightful machine to fly and I enjoyed putting it through its paces to see what it could do. It was rumored the Tempest V could not recover from a certain type of spin. I had to find out! I went up over 30,000 feet to give myself plenty of time to bail out if necessary. I then proceeded to perform every type of spin I could think of, including letting them really wind up. However, there was never a problem recovering from any of them, so we knew it was a safe and well-designed aircraft.

Sometime around the middle of October 1945, I was leading a practice flight of four. I peeled off as if to attack a ground target when my controls locked and I was unable to move the stick. I was preparing to take a walk when the control column freed up. I pulled back up and handed over command to my number two as I headed back to our airfield. It was a very pleasant day and I was cruising at 400 mph at a good altitude when my engine quit cold. I figured I had enough speed and altitude to reach our airfield so I reported I was coming in dead stick and to clear the way for me. There was no reply from the tower. However, the alert ground crew saw me coming in with my prop just wind milling and drove over to pick me up after I stopped. It turned out the controllers in the tower were so involved in some sort of black market haggling they had not heard my call. The amazing thing was that after going over the machine they could not find any reason why the controls would lock up. We assumed that someone left a spanner (English for wrench) in the fuselage, but if so they removed it and never reported it. Once again the hand of the Lord was on me in an incredible way, and saved me from a very dangerous situation. Call it what you may, knowing the engine would quit, our Lord Jesus Christ had prepared the way for me to land safely at my base. Think about it.

On 10 November 1945 I was asked to pick up a Tempest V from a base in France. It was quite late in the day when I arrived and got off the ground so I decided to stop off at Brussels, Belgium, for the night. In the morning it was a beautiful clear day. After leaving the Brussels area cruising at 400 mph I went down to treetop level and flew across Holland and Germany to our base at Lübeck on the Baltic and landed. I was amused later when I heard the Russians had complained I was flying over their territory. They assumed I would continue on across the border, very close to our airfield, which of course I did not.

A Tribute to Group Captain Johnny Johnson

41 Squadron was one of the squadrons under Group Captain Johnny Johnson. He was a top-flight pilot, obviously, since he not only survived the Battle of Britain, but also the following years of the war. He was a wise and intrepid leader. It was my considerable honor to carry out a

number of special missions for him in the months immediately following the end of the war in Europe.

Johnny Johnson was a man who pressed life and experience to its limits. In Europe, he owned fast cars, which he drove with abandon. One of them was a large supercharged Mercedes Benz which he drove from Lübeck to Hamburg in Germany, a distance of 60 miles in half an hour. You figure out what his average speed was! One day our Adjutant was riding with Johnny Johnson who was driving excessively fast as usual; not daring to look, the Adjutant put hands over his eyes. When he peeked through the fingers of his hands, he thought he was about to die. He could see one of the wheels speeding away across a nearby field. Fortunately it turned out to be the spare, which was stored on the side of that powerful vehicle.

One day Johnny Johnson gave me a lift from the airfield to the mess. He drove at such a speed he was taking curves on two wheels. Needless to say, as the Adjutant, it was the last time I accepted a lift from him!

One of my best and most treasured recollections of Johnny Johnson was when he challenged me to a friendly dogfight. We took off in formation and climbed to about 15,000 feet where we turned away from each other. When we were a few miles apart we turned and faced each other. From then on we twisted and turned and climbed and dived and rolled around, pressing our Spitfires and ourselves to the very limits. It was truly exhilarating. However, try as I would I could not get the better of him. But then, try as he would he could not get the better of me. So it was a draw and we each had a healthy respect for each other's flying ability. He was the only one I could not defeat in a dogfight.

Johnny and I had the same method of fighting; that is to press machine and body to the very limits. When it came to real life fighting with the enemy fighters I didn't mess around, I finished them off quickly.

Johnny Johnson was a fine man who lived life to the fullest and I understand lived to a ripe old age. If he now has heavenly wings I am sure he must be flying rings around all the other angels!

Reading on the Thames River west of London

After the war I attended the RAF Flight Instructor course to become an Air Force flying instructor. Now I was teaching students to fly the Tiger Moths, a low powered biplane with no brakes, no radio, no flaps, no nothing! I had come full circle, I was back where I started but I was still flying.

One of the maneuvers students must perfect with frequent practice is forced landings. There was a specific area for this purpose. I would have the student fly into the area and then, without warning, I would pull the throttle back. The student would then have to go through specific maneuvers, judge the distance and glide down to the field. It was springtime and the approach to the field was covered by a large area of rhododendrons of all colors. It was so beautiful; I would sit there and admire the view as long as the student performed his tasks properly.

In July 1946, I was demobilized.

***

John F. Wilkinson

41 Squadron Badge Hearldry

On the front cover

A red double-armed cross on a white background adapted from the coat of arms of the city of Saint Omer, France and approved by His Majesty King George VI in February 1937. The badge originated from the squadron's association with St Omer, which, in 1916, was the unit's first overseas base during World War I.

Motto:

Seek and Destroy

No. 41 Squadron of the Royal Air Force is based at Coningsby, Lincolnshire. The Squadron is one of the oldest RAF squadrons in existence, celebrating its 100th Anniversary in 2016.

ABOUT THE AUTHOR

FOLLOWING THE WAR, John acquired passage as the only passenger on the SS Teespool, a freighter that had no destination of record. They got underway from Wales on New Year's Eve 1948. Following a brief stay in Cuba, the Teespool crossed the Gulf of Mexico and went up the Mississippi to New Orleans where John left the ship on February 14, 1949, took a bus to Kansas City and attended Unity Institute for 5 Years.

John married Jane Sessions on April 18, 1952. He served as an Assistant Pastor in Oakland, California before moving to Billings, Montana where his daughter Heidi Jane was born. In Billings he served as a church Pastor before moving, by request, to Vancouver, Canada. In 1957, he became a naturalized citizen of the United States.

John left the pastoring field in favor of a career in business. In the 60's he and Jane parted. During the following years, he worked as an innovator in the field of electronic switching mechanisms, computer and photographic equipment. He produced several inventions during that time.

In 1968 John took a trip to England to visit his homeland and two sisters. Following a Christian service John was born-again, as described in the Gospel of John, chapter three. He was forever transformed - the greatest event in his life, during which he developed a personal relationship with Jesus Christ and the Holy Spirit who gave him the assurance of eternal life with Him in Heaven and the gift of speaking in Tongues. This is available to all who want to know and have the courage to ask (Revelation 3:20).

In 1974, he met and married fellow Englishman Joyce Simmons who had been a WWII RAF aircraft plotter. With yet another company John installed computers, a CAD/CAM system, wrote programs and maintained equipment. With the advent of personal computers, he merged data from two dissimilar systems and was responsible for programming and maintenance of a hundred computers. John retired from the work force at 68.

After retiring, he and Joyce moved to Redding, California where they lived for ten years. They then moved to Oak Harbor on Whidbey Island for a preferable climate. When it became necessary to move closer to medical facilities John rented a small condominium in Panorama City where a neurologist diagnosed Joyce with Alzheimer's disease.

Joyce's older son convinced John that he and Joyce should move to Spearfish, South Dakota. It was a wise move. They bought a town house in Spearfish in a friendly community near family. Later it became necessary to move Joyce into Edgewood Vista, an assisted living and memory care facility in Spearfish for Alzheimer patients.

* * *